T0406287

Springer Complexity

Springer Complexity is an interdisciplinary program publishing the best research and academic-level teaching on both fundamental and applied aspects of complex systems – cutting across all traditional disciplines of the natural and life sciences, engineering, economics, medicine, neuroscience, social and computer science.

Complex Systems are systems that comprise many interacting parts with the ability to generate a new quality of macroscopic collective behavior the manifestations of which are the spontaneous formation of distinctive temporal, spatial or functional structures. Models of such systems can be successfully mapped onto quite diverse "real-life" situations like the climate, the coherent emission of light from lasers, chemical reaction-diffusion systems, biological cellular networks, the dynamics of stock markets and of the internet, earthquake statistics and prediction, freeway traffic, the human brain, or the formation of opinions in social systems, to name just some of the popular applications.

Although their scope and methodologies overlap somewhat, one can distinguish the following main concepts and tools: self-organization, nonlinear dynamics, synergetics, turbulence, dynamical systems, catastrophes, instabilities, stochastic processes, chaos, graphs and networks, cellular automata, adaptive systems, genetic algorithms and computational intelligence.

The two major book publication platforms of the Springer Complexity program are the monograph series "Understanding Complex Systems" focusing on the various applications of complexity, and the "Springer Series in Synergetics", which is devoted to the quantitative theoretical and methodological foundations. In addition to the books in these two core series, the program also incorporates individual titles ranging from textbooks to major reference works.

Understanding Complex Systems

Future scientific and technological developments in many fields will necessarily depend upon coming to grips with complex systems. Such systems are complex in both their composition – typically many different kinds of components interacting simultaneously and nonlinearly with each other and their environments on multiple levels – and in the rich diversity of behavior of which they are capable.

The Springer Series in Understanding Complex Systems series (UCS) promotes new strategies and paradigms for understanding and realizing applications of complex systems research in a wide variety of fields and endeavors. UCS is explicitly transdisciplinary. It has three main goals: First, to elaborate the concepts, methods and tools of complex systems at all levels of description and in all scientific fields, especially newly emerging areas within the life, social, behavioral, economic, neuro- and cognitive sciences (and derivatives thereof); second, to encourage novel applications of these ideas in various fields of engineering and computation such as robotics, nano-technology and informatics; third, to provide a single forum within which commonalities and differences in the workings of complex systems may be discerned, hence leading to deeper insight and understanding.

UCS will publish monographs, lecture notes and selected edited contributions aimed at communicating new findings to a large multidisciplinary audience.

Cyrille Bertelle · Gérard H.E. Duchamp ·
Hakima Kadri-Dahmani (Eds.)

Complex Systems and Self-organization Modelling

Professor Cyrille Bertelle
LITIS (EA 4051)
UFR Sciences et Techniques
25 rue Ph. Lebon - BP 540
76058 Le Havre Cedex
France
cyrille.bertelle@univ-lehavre.fr

Dr. Hakima Kadri Dahmani
Laboratoire d'Informatique de l'université Paris Nord
Institut Galilée
99 avenue J.B. Clément
93430 Villetaneuse
France
hkd@lipn.univ-paris13.fr

Professor Gérard H.E. Duchamp
LIPN - UMR CNRS 7030
University of Paris 13
99 avenue Jean-Baptiste Clément
93400 Villetaneuse
France
ghed@lipn.univ-paris13.fr

ISBN: 978-3-540-88072-1 e-ISBN: 978-3-540-88073-8

DOI 10.1007/978-3-540-88073-8

Understanding Complex Systems ISSN: 1860-0832

Library of Congress Control Number: 2008936468

Cover design: WMXDesign GmbH

Printed on acid-free paper

9 8 7 6 5 4 3 2 1

springer.com

Preface

The concern of this book is the use of emergent computing and self-organization modelling within various applications of complex systems. We focus our attention both on the innovative concepts and implementations in order to model self-organizations, but also on the relevant applicative domains in which they can be used efficiently.

First part deals with general modelling and methodology as conceptual approaches for complex systems description. An introductive chapter by Michel Cotsaftis entitled "A Passage to Complex Systems", treats the notion of "Complex Systems" in opposition to that of a "Complicated System". This can be, he claims, comprehended immediately from the latin roots as "Complex" comes from "cum plexus" (tied up with) whereas "complicated" originates from "cum pliare" (piled up with). The paper is a wide and rich dissertation with elements of history (of the technical developement of mankind) with its recents steps : mechanist, quantum and relativistic points of view. Then, the need for a "passage" is illustrated by the discussion, with tools borrowed from functional analysis, of a typical parametric differential system. The last and conclusive parts give tracks for the study of Complex Systems, in particular one can hope to pass to quantitative study and control of complex systems even if one has to consent a "larger intelligence delegation" to them (as announced in the introduction) by using and developing tools already present in dissipative Physics and in Mathematical functional analysis and fixed point theorems, for instance. This "passage" is followed by a wide bibliography of more than 90 entries. The (non hasty) reader is invited to read this deep and far reaching account before browsing through the book.

The chapter, "Holistic Metrics, a Trial on Interpreting Complex Systems" by J. M. Feliz-Teixeira et al., proposes a simple and original method for estimating or characterize the behaviour of complex systems, in particular when these are being studied throughout simulation. The originality of the chapter lies in the fact that the time/observable space is replaced by the corresponding variable/observable space (as one does for Wavelet Transforms and in Quantum Mechanics). Next chapter, “Different Goals in Multiscale Simulations and How to Reach Them” by P. Tranouez et al., summarizes the works of the authors on multiscale programs, mainly simulations. They present methods for handling the different scales, with maintaining a summary, using an environmental marker introducing a history in the data and finally using knowledge on the behaviour of the different scales to handle them at the same time. “Invariant Manifolds in Complex Systems” by J.-M. Ginoux et al. shows how to locate, in a general dynamical system (on a 2,3 dimensional variety) remarkable subsets which are flow-invariant. Part I ends with a chapter by Z. Odibat et al. entitled “Application of Homotopy Perturbation Method for Ecosystems Modelling” (HPM). HPM is one of the new methods belonging ranking as one of the perturbation methods. The attention of the reader is focused on the generation of the decomposition steps to build a solver using the HPM method. Concrete solvers for prey-predator systems involving 2 or 3 populations are computed and a special attention is paid on implementation aspects.

Second part deals with swarm intelligence and neuronal learning. We focus our attention here on how implement self-organization processes linked to applicative problems. Both swarm intelligence and neuronal learning give some ways to drive the whole system, respecting its complex structure. F. Ghezail et al. use one of the most efficient swarm intelligence processes, ant colonies method, to solve a multi-objective optimization problem. J. Franzolini et al. present a very promising new approach based on swarm intelligence, immune network systems. They give detailed explanation on the biological metaphor and accurate simulation results. The last chapter of this part, by D.A. El-Kebbe et al., deals with the modelling of complex clustering tasks involved in cellular manufacturing, using neural networks. On the basis of Kohonen's self-organizing maps, they introduce Fuzzy Adaptive Resonance Theory (ART) networks to claim on their efficiency to obtain consistent clustering results.

Third part entitled “Socio-Environmental Complex Modelling and Territorial Intelligence”, deals with the complexity of systems where space is fundamentally the center of the interaction network. This space interacts on the one hand, with human themselves or their pre-defined or emergent organizations and on the other hand within natural processes, based on living entities inside ecosystems or also on physical features (like in the complex multi-scale phenomena leading to cliff collapse hazards described by Anne Duperret et al.). In the first case, we focus on geographical information systems (GIS) where humans are now able to notify, with an accuracy of location, the material based on their own organization. Even if these GIS contitute an impressive database in static way at a fixed time, they are still not able to restitute the complexity of the human organization dynamics and we propose in this book some research developments to lead their evolution toward their inherent complexity. H. Kadri-Dahmani et al. study the emergent properties from the GIS updating propagation process over an interactive network; R. Ghnemat et al. focus on the necessity of mixing GIS with active processes called agents which are able to generate emergent organization from basic simple rules like in Schelling's segregation model; D. Provitolo proposes a methodology deeply inspired from the complexity concepts, for modelling risk and catastrophe systems within dynamical systems; G. Prevost et al. propose an effective methodology, based on adaptative processes, to mix the two majors classes of simulation: differential approach and individual-based approach. Through the unavoidable expression of the complexity expressed in these different contributions, we can feel how the Complexity Science renovates the modelling approaches, respecting and highlighting the fundamental and classical methods by the "cum-plexus" combination of them to express the whole system complexity, more than by the addition of a long list of complicated scattered sub-systems.

Fourth part deals with emotion modelling within the cognitive processes as the result of complex processes. The general purpose here is to try to give some formal description to better understand the complex features involved in the essential emotion–cognition–action interaction. Decision making is one of the result of this interaction: K. Mahboub et al. study and propose a model to mix in a complex way the emotional aspects in some player choices. In a second paper, S. Baudic et al. propose a relevant approach leading to confront theory and clinical practice to better improve the knowledge of emotion and its interaction with memory (with practical illustration based on Alzheimer's disease) and with cognition (through the fear behaviour). Therapeutic applications can then be implemented from this methodology.

Fifth part deals with simulation and production systems. In that field, Complexity Science gives a new way to model the engineering process involved in some productions systems dealing with the management of a great number of components and dimensions in multi-representation and multi-scale description. The contribution of B. Kausch et al. deals with this complex process, applied to chemical engineering, using Petri nets modelling. The contribution of G. Giulioni claims that self-organization phenomena and complexity theory is a relevant way to model economic reality. This study proposes a model based on the economic result of a large number of firms based on the evolution of capital and the dynamics of productivity. The discussion from output results enlights the emergence of attractors on the aspects of limit cycles and possible transition to equilibrium. The contribution of A. Dumbuya et al. deals with the complexity of traffic interaction and the development of a driver model based on neural networks. The goal is to improve the behavioural intelligence and realism in driving simulation scenarios.

This book is the outcome of a workshop meeting within ESM 2006 (Eurosis), held in Toulouse (France) in october 2006, under the efficient organization of Philippe Geril that we would like to thank here.

Le Havre & Paris, France,
April 2008

Cyrille Bertelle
Gérard H.E. Duchamp
Hakima Kadri-Dahmani

Contents

Part I Complex system modelling and methodology

A Passage to Complex Systems
Michel Cotsaftis 3

Holistic Metrics, a Trial on Interpreting Complex Systems
J. Manuel Feliz-Teixeira and António E. S. Carvalho Brito 21

Different Goals in Multiscale Simulations and How to Reach Them
Pierrick Tranouez and Antoine Dutot 29

Invariant Manifolds of Complex Systems
Jean-Marc Ginoux and Bruno Rosseto 41

Application of Homotopy Perturbation Method for Ecosystems Modelling
Zaid Odibat and Cyrille Bertelle 51

Part II Swarm intelligence and neuronal learning

Multi Objective Optimization Using Ant Colonies
Feïza Ghezail, Henri Pierreval, and Sonia Hajri-Gabouj 65

Self-Organization in an Artificial Immune Network System
Julien Franzolini and Damien Olivier 71

On Adapting Neural Network to Cellular Manufacturing
Dania A. El-Kebbe and Christoph Danne 83

Part III Socio-environmental complex modelling and territorial intelligence

The Evolution Process of Geographical Database within Self-Organized Topological Propagation Area
Hakima Kadri-Dahmani, Cyrille Bertelle, Gérard H.E. Duchamp, and Aomar Osmani 97

Self-Organization Simulation over Geographical Information Systems Based on Multi-Agent Platform
Rawan Ghnemat, Cyrille Bertelle, and Gérard H.E. Duchamp 107

Cliff Collapse Hazards Spatio-Temporal Modelling through GIS: from Parameters Determination to Multi-scale Approach
Anne Duperret, Cyrille Bertelle, and Pierre Laville 117

Structural and Dynamical Complexities of Risk and Catastrophe Systems: an Approach by System Dynamics Modelling
Damienne Provitolo 129

Detection and Reification of Emerging Dynamical Ecosystems from Interaction Networks
Guillaume Prévost and Cyrille Bertelle 139

Part IV Emotion and cognition modelling

Simulation of Emotional Processes in Decision Making
Karim Mahboub and Véronique Jay 165

Emotions: Theoretical Models and Clinical Implications
Sophie Baudic and Gérard H. E. Duchamp 177

Part V Production systems and simulation

Complex Systems Dynamics in an Economic Model with Mean Field Interactions
Gianfranco Giulioni 189

Complexity of Traffic Interactions: Improving Behavioural Intelligence in Driving Simulation Scenarios
Abs Dumbuya, Anna Booth, Nick Reed, Andrew Kirkham, Toby Philpott, John Zhao, and Robert Wood 201

An Integrative Simulation Model for Project Management in Chemical Process Engineering
Bernhard Kausch, Nicole Schneider, Morten Grandt, and Christopher Schlick 211

Index 233

Part I

Complex system modelling and methodology

A Passage to Complex Systems

Michel Cotsaftis

LACSC - ECE
53 rue de Grenelle
75007 Paris, France
mcot@ece.fr

Summary. Complex systems are the new scientific frontier which was emerging in the past decades with advance of modern technology and the study of new parametric domains in natural systems. An important challenge is, contrary to classical systems studied so far, the great difficulty in predicting their future behaviour from an initial instant as by their very structure the interactions strength between system components is shielding completely their specific individual features. So these systems are a counterexample to reductionism so strongly influential in Science with Cartesian method only valid for complicated systems. Whether complex systems are obeying strict laws like classical systems is still unclear, but it is however possible today to develop methods which allow to handle some dynamical properties of such system. They should comply with representing system self organization when passing from complicated to complex, which rests upon the new paradigm of passing from classical trajectory space to more abstract trajectory manifolds associated to natural system invariants characterizing complex system dynamics. So they are basically of qualitative nature, independent of system state space dimension and, because of generic impreciseness, privileging robustness to compensate for not well known system parameter and functional variations. This points toward the importance of control approach for a complex system, the more as for industrial applications there is now evidence that transforming a complicated man made system into a complex one is extremely beneficial for overall performance improvement. But this requires larger intelligence delegation to the system, and a well defined control law should be set so that a complex system described in very general terms can behave in a prescribed way. The method is to use the notion of equivalence class within which the system is forced to stay by action of the control law constructed in explicit terms from mathematical (and even approximate) representation of system dynamics.

1 Introduction

After observation of natural most visible phenomena for many centuries, Man realized that they are depending on an order which He was then willing to discover. From hypotheses He made as simple and acceptable as possible, He constructed laws representing these phenomena and allowing Him to predict them, starting from the most elementary ones corresponding to simple systems defined as being easily isolated in their observation and their evolution, and later extending them to complicated systems by reductionism. This millennium long quest produced adapted representation of the universe, with mainly a "classical" one for human size phenomena, corrected by "quantum" effects at atomic and sub-atomic infinitely small size and "relativistic" ones at galactic and extra-galactic infinitely large ones. However a new situation gradually emerged from the extraordinary advance in modern technology which took place soon after World War II, both with natural (including living organisms) and man made systems. Recent and finer observations of first ones have shown the existence of another very broad class of systems gathering a very large number of heterogeneous components and characterized by the extremely strong level of their mutual interactions with considerable impact on final system output. Amongst them the most advanced living systems are unravelling with great pain their extraordinarily intricate structure allowing them to perform highly advanced tasks for their survival. In the same way man made systems have also reached a level with individual very highly performing components, and the acute problem is today to take full advantage of their specific capability. These new coming systems called complex (from Latin origin "cum plexus" : tied up with) systems are no longer reducible to simple systems like complicated (from Latin origin "cum pliare" : piled up with) ones by Descartes method. The difficulty comes from observation that usual splitting of real world dynamics into mechanical and thermodynamic representations is not enough to handle these structures staying in between as their global response is not predictable from strict mechanical component behaviour but is not approachable either by thermodynamic global analysis as convective effects are still important. New ways have to be developed for proper analysis of their dynamics which do not come out from just addition of the ones of their components, and the research of final system behaviour is, due to importance of nonlinearities, generally outside the range of application of classical methods. This is understandable inasmuch as from observations in many natural domains complex system often reaches its stage after exhibiting a series of branching along which it was bifurcating toward a new global state the features of which are not usually amenable to a simple local study, being remembered that the branching phenomenon is resting upon a full nonlinear and global behaviour. There exists another important class of systems composed of agents with definite properties for which complex stage is reached via the firing of interactions between the agents. In this case it is usually said that there is emergence of a new behaviour even if the term is not clearly defined yet[1]. However, despite still many divergent definitions of "complex system"[2] a more and more accepted one is that a system going into complex stage becomes self organized[3]. A fundamental and immediate consequence is that the determination of system dynamics now requires manipulation of less degrees of freedom than the system has initially. For man made systems, this means that their control in complex stage is realized when acting on a restricted number of degrees of freedom, the other ones being taken care of by internal reorganization. More precisely there is impossibility to control all degrees of freedom of a system from outside when interaction strengths between its components are above some threshold value, and dynamical effects are still effective (so the threshold value is below the ultimate "statistical" threshold above which the system looses its dynamics and becomes random, entering the domain of thermodynamics). A reason is that when crossing the threshold value from below, characteristic internal interaction time between interested components becomes shorter than (fixed) characteristic cascading time from outer source. Thus internal power flux overpasses outer cascading power flux and drives in turn interested components dynamics up to last thermodynamic stage where characteristic internal interaction time becomes so short that internal system dynamics are replaceable by ergodic ones. Globally, in the same way as there are three states of matter (solid, liquid, gas), there are also three stages of system structure (simple, complicated, complex). Each stage exhibits its own features and corresponding organization which, for the first two stages at least, is manifested by physical laws. In mathematical terms, one is gradually passing when increasing interaction strength between system components from all microscopic system invariants (the 6 N initial positions and velocities) in mechanical stage to more and more global ones when entering complex stage up to the last thermodynamic stage where the only system invariant is energy (and sometimes momentum). This has been observed when numerically integrating differential systems and seeing that trajectories are filling in denser way larger and larger domains in phase space with existence of chaos, invariant manifolds and strange attractors[4] when increasing the value of coupling parameter. From this very rapid description there is a double challenge in understanding the behaviour of complex stage by building new adapted tools, and, because of their consequences, by developing new methods for application to modern technology. A more advanced step is also to understand why natural complex systems are appearing from their complicated stage. Some elements of these questions will be considered in the following.

2 Methods of Approach

For a large class of natural systems, complex stage is thus usually occurring after branching process leading to internal reorganization necessarily compatible with system boundary conditions, basically a nonlinear phenomenon. Various methods do exist to deal with both in Applied Mathematics and in Control methods. In first group, where interest is in analyzing system properties as it stands, results on "chaotic" state show that the later represents the general case of non linear non integrable systems[5], reached for high enough value of (nonlinear) coupling parameter. In second group another dimension is added by playing on adapted parameters to force system behaviour to a fixed one. Despite its specific orientation it will appear later that this way of approach when properly amended is more convenient to provide the framework for studying systems in complex stage. Are belonging to this group extensive new control methods often (improperly) called "intelligent"[6], supposed to give systems the ability to behave in a much flexible and appropriate way. However these analyses, aside unsolved stability and robustness problems[7], still postulate that system trajectory can be followed as in classical mechanical case, and be acted upon by appropriate means. In present case on the contrary, the very strong interaction between components in natural systems induces as observed in experiments a wandering of trajectory which becomes indistinguishable from neighbouring ones[8], and only manifolds can be identified and separated[9]. So even if it could be tracked, specific system trajectory cannot be modified by action at this level because there is no information content from system point of view, as already well known in Thermodynamics[10]. Similar situation occurs in modern technology applications where it is no longer possible to track system trajectory. Only dynamic invariant manifolds can be associated to tasks to give now systems the possibility to decide their trajectory for a fixed task assignment (for which there exists in general many allowed trajectories). Whether from natural systems or from models, in both cases there is a shift to a situation where the mathematical structure generates a manifold instead of a single trajectory, now needed for fulfilling technical requirements in task execution under imposed (and often tight) economic constraints. This already very important qualitative jump in the approach of highly nonlinear systems requires proper tools for being correctly handled. So to analyze complex systems applicable methods should adapt to the new manifold paradigm and the first step is to consider system trajectory as a whole $x(.)$ instead of the set $[x(t_1),\ x(t_2)\ ...\ x(t_n)]$ for the time instants $[t_1,\ t_2\ ...\ t_n]$ in usual mechanistic approach. As a consequence, handling a (complete) trajectory as elementary unit requires a framework where this is possible, and the convenient one is Functional Analysis. Then the problem of complex system dynamics can be reformulated in the following way. Consider the finite dimensional nonlinear and time dependent systems

$$\frac{dx_s}{dt} = F_s(x_s(t), u(t), d(t), t) \tag{1}$$

where $F_s(.,.,.,.) : \mathbb{R}^n \times \mathbb{R}^m \times \mathbb{R}^p \times \mathbb{R}^1 \mapsto \mathbb{R}^n$ is a C^1 function of its first two arguments, $x_s(t)$ the system state, $u(t)$ the control input, and $d(t)$ the disturbance acting upon the system. In full generality the control input $u(t)$ can be either a parameter which can be manipulated by operator action in man made system or more generally an acting parameter on the system from its environment to the variation of which it is intended to study the sensitivity. To proceed, this equation will be considered as a generic one with now $u(.) \in \mathcal{U}$ and $d(.) \in \mathcal{D}$, where $\mathcal{U}$ and $\mathcal{D}$ are two function spaces to be defined in compatibility with the problem, for instance $\mathcal{L}^p$, $\mathcal{W}^p_n$, $\mathcal{M}^p_n$, respectively Lebesgue, Sobolev and Marcenkievitch-Besicovitch spaces[11] related to useful and global physical properties such as energy and/or power boundedness and smoothness. Now for $u(.)$ and $d(.)$ in their definition spaces, eqn(1) produces a solution $x_s(.)$ which generates a manifold $\mathcal{E}$ and the problem is now to analyze the partitioning of $\mathcal{U}$ and $\mathcal{D}$ corresponding to the different (normed) spaces $\mathcal{S}$ within which $\mathcal{E}$ can be embedded. When $\mathcal{S}$ is $\mathcal{M}_n^2$ for instance simple stability property is immediately recovered. The base method to express this property is the use of fixed point theorem in its various representations [12]. The generality of this approach stems to the fact that all stability and embedding methods written so far since pioneering work of Lyapounov[13] and Poincaré[14] are alternate expressions of this base property[15]. The problem is easily formulated when there exists a functional bounding $F_s(.,.,.,.)$ in norm. For instance, a usual bound (related to Caratheodory condition in Thermodynamics) is in the form of generalized Lipschitz inequality[16]

$$||F_s(x,u,d,t) - F_s(x',u',d',t')|| \leq$$
$$L_s(t)||x-x'||^{\lambda_s} + L_1(t)(||u-u'||_\infty + ||d-d'||_\infty \tag{2}$$

Then by substitution one gets for $x_s(t)$ the bound

$$||x_s(t)|| \leq ||x_s(t_0)|| + \int_{t_0}^{t} \Big[L_s(t')||x_s(t')||^{\lambda_s} + R_1(t')\Big]dt' \tag{3}$$

with $R_1(t) = L_1(t)\Big[||u(t)||_\infty + ||d(t)||_\infty\Big]$ and when solving for $x_s(t)$

$$\int_{t_0}^{t} R_1(t')dt' \geq ||x_s(t)||\ {}_2F_1\Big(1, \lambda_s^{-1}, 1+\lambda_s^{-1}; -Sup\Big[\frac{L_s(t)}{R_1(t)}\Big]||x_s(t)||^{\lambda_s}\Big) \tag{4}$$

with ${}_2F_1(\alpha,\beta,\gamma;z)$ the hypergeometric function[17]. So there is a fixed point $x_s(t) \in \mathcal{L}_\infty$ for $u, d \in \mathcal{L}_\infty$ exhibiting simple stability property. The result extends to more general non decreasing bounding function $g(||x-x'||)$ instead of polynomial one in eqn(2) [18]. A further step is obtained when representing eqn(1) close to a solution $x_{s_0}(t)$ obtained for a specific input $u_0(t)$ by

$$\frac{d\Delta x_s}{dt} = A\Delta x_s + B\Delta u + \eta \tag{5}$$

with $x_s(t) = x_{s_0}(t) + \Delta x_s$ and

$$u(t) = u_0(t) + K\Delta x_s + \Delta u \tag{6}$$

splitting apart linear terms in Δx_s and Δu, and where is standing for all the other terms. If there exists a bound $|\eta| \leq \rho(|\Delta x_s|, |\Delta u|, t)$ it is possible to find the functional form of u so that $\Delta x_s(t)$ belong to a prescribed function space. For instance with the expression[19]

$$\Delta u = \frac{\rho(B^T B)^{-1} B^T P \Delta x_s}{\alpha ||P \Delta x_s|| + \epsilon f(.)} \tag{7}$$

where K is supposed to be such that $A^T P + PA + \frac{dP}{dt} = -Q$ has a solution (P, Q) positive definite, and $f(.)$ to be determined, one gets the bounding equation for Δx_s

$$\frac{dX}{dt} = -\lambda_{min}(Q)X + 2\frac{\epsilon}{\alpha}\rho.f(.) \tag{8}$$

For given functional dependence $\rho(|\Delta x_s|, |\Delta u|, t)$, the problem can now be stated as the determination of the functional dependence $f(.)$ for which $X(t)$ exhibits specific behaviour in definition interval (in general for large t). Differently said, this is researching correspondence between function spaces $\mathcal{X}$ and $\mathcal{F}$, with $X(.) \in \mathcal{X}$ and $f(.) \in \mathcal{F}$, so that for given $\rho(.,.)$ $\mathcal{X} \subset \mathcal{S}$ initially fixed. In present case more specifically, eqn(2) becomes $\rho(X, \Delta u, t) \leq a(t) + b(t)X^{p/\mu}$. Substitution theorems[20] for Sobolev spaces gives $\rho(X, ., t) \in \mathcal{W}_1^p$ if $a(.) \in \mathcal{L}^p$, $b(.) \in \mathcal{L}^q$ and $1/p = 1/q + \mu/p$, and application of Hölder inequality to eqn(8) gives $X(.) \in W_1^r$ if $f(.) = kX^s$ so that $1/r = 1/p + 1/s$. In physical terms eqn(8) expresses that, under the adverse actions of the attractive harmonic potential in first term of the right hand side and of the globally repulsive second one from the nonlinear terms, its solution belongs also to a Sobolev space $\mathcal{S}$ defined by previous relations between index of the various initial Sobolev spaces. Note that when $a(.) = 0$, eqn(8) becomes a Bernouilli equation with solution

$$X(t) = \frac{X_0 exp(-\lambda_{min}(Q)t)}{\left[1 - \frac{2\epsilon(p/\mu+s-1)X_0^{(p/\mu+s-1)}}{\alpha} \int_{t_0}^t b(t')exp(-(p/\mu + s - 1)t')dt'\right]^{\frac{1}{p/\mu+s-1}}} \tag{9}$$

with $X_0 = X(t_0)$, exhibiting a non exponential but asymptotic time dependence for large t when there is no zero in the denominator defining the attraction domain of eqn(1) in terms of actual parameters and initial conditions. Importantly the result of eqn(9) shows that there is an equivalence robustness class for all equations having the same bounding equation as the proposed approach is not focusing on a specific and single eqn(1) but on a class here defined with few parameters. The interesting point is the role of the attractive harmonic potential included in the expression of $u(.)$ in eqn(6) defining the attraction class in $\mathcal{S}$. More generally this suggests the very simple picture of a "test" of eqn(1) on a prescribed space $\mathcal{S}$ by a set of harmonic springs over a base set of $\mathcal{S}$. Evidently if the smallest of the springs is found to be attractive for actual parameters $u(.)$ and $d(.)$ the embedding is realized in $\mathcal{S}$. So the embedding is solely controlled by the sign of the smallest spring, which is an extremely weak and clearly identified knowledge about the system under study. This opens on the application of spectral methods which appear to be particularly powerful because they are linking a evident physical meaning (the power flow) with a well defined and operating method to construct a fixed point in the target space $\mathcal{S}$[15]. Obviously the number of dimensions of the initial system is irrelevant as long as only the smallest spring force (the smallest eigenvalue of system equation in space $\mathcal{S}$) is required. In this sense such result is far more efficient than usual Lyapounov method which is a limited algebraic approach to the problem. Another element coming out of previous result is the fact that present approach is particularly well tailored for handling the basic and difficult problem of equivalence, especially asymptotic equivalence where system dynamics reduce for large time to a restricted manifold, and sometimes a finite one even if initial system dimension is infinite as for instance for turbulent flow in Fluids dynamics[21]. So asymptotic analysis is also a very powerful tool for studying complex systems, especially when they belong to the class of reducible ones. Such systems are defined by the fact that the bifurcation phenomenon which generates the branching toward more complex structure is produced by effects with characteristic time and space scales extremely different (and much smaller) from base system ones. So the system can be split into large and small components the dynamics of which maintain system global structure from the first when interacting with the second in charge of dissipation because they have lost their phase correlations, hence the name of dissipative structures[22]. Due to smallness and indistinguishability the initial values of small components are obeying central limit theorem and are distributed according to a Gaussian. However, it is not possible to neglect at this stage their dynamics which can on a (long) time compared to their own time scale act significantly on large components dynamics, and the usual Chandrasekhar model[23] does not apply here. Small components dynamics can be asymptotically solved on (long) large components time scale, and injected in large components ones. Then system dynamics are still described by large components, but modified by small components action[24].

3 Mastering Complex Systems

In previous part a few methods have been defined to elucidate the new challenge represented by the understanding of complex systems dynamics, mainly because of their internal self-organization shielding the access to their full dynamics as for previous simple and complicated systems. Here by choosing an approach based on how to act onto the system much more than on a simple description, it has been seen that advantage of this feature can be taken by abandoning initial and too much demanding mechanistic point of view for a more global one where outer action is limited to acceptable one for the system fixed by its actual internal organization state. It is immediately expectable that by reducing outer action to this acceptable one, one could expect the system to produce its best performance in some sense because it will naturally show less dependence on outer environment. An elementary illustration of this approach is given by the way dogs are acting on a herd of cattle in the meadows. If there are n animals wandering around, they represent a system with $2n$ degrees of freedom ($3n$ when counting their orientation). Clearly the dog understands that it is hopeless with his poor 2 degrees of freedom to control all the animals, so his first action is to gather all the animals so that by being close enough they have strong enough interactions transforming their initial complicated system into a complex one with dramatically less degrees of freedom, in fact only two like himself. Then he can control perfectly well the herd as easily observed. The astonishing fact is that dogs are knowing what to do (and they even refuse to do anything with animals unable to go into this complex stage) but today engineers are not yet able to proceed in similar way with their own constructions. This is an immense challenge industrial civilisation is facing today justifying if any the needs to study these complex systems. This has to be put in huge historical perspective of human kind, where first men understood that they had to extend the action of their hand by more and more adapted tools. This was the first step of a delegation from human operator to executing tool, after to executing machine, which is now entering a new and unprecedented step because of complexity barrier, where contrary to previous steps, system trajectory now escapes from human operator who is confined in a supervision role. As indicated earlier, the new phenomenon of self organization does not allow to split apart single system trajectory as before, and forces to consider only manifolds which are the only elements accessible to action from environment. So the system should now be given the way to create its own trajectory contrary to previous situations where it was controlled to follow a predetermined one. This requires a special "intelligence" delegation which, as a consequence, implies the possibility for the system to manipulate information flux in parallel to usual power flux solely manipulated in previous steps. Strikingly Nature has been facing this issue a few billion years ago when cells with DNA "memory" molecules have emerged from primitive environment. They exhibit the main features engineers try today to imbed in their own constructions, mainly a very high degree of robustness resulting from massive parallelism and high redundancy. Though extremely difficult to understand, their high degree of accomplishment especially in the interplay between power and information fluxes may provide interesting guidelines for technical present problem. Classical control problem[25] with typical control loops guaranteeing convergence of system output toward a prescribed trajectory fixed elsewhere, shifts to another one where the system, from only task prescription, has to generate its own trajectory in the manifold of realizable ones. A specific type of internal organisation has to be set for this purpose which not only gives the system necessary knowledge of outside world, but also integrates its new features at system level[26]. In other words, it is not possible to continue classical control line by adding ingredients extending previous results to new intelligent task control. Another type of demand is emerging when mathematically passing from space time local trajectory control to more global manifold control based on giving the system its own intelligence with its own capacities rather than usual dump of outer operator intelligence into an unfit structure. The question of selecting appropriate information for task accomplishment and linking it to system dynamics has also to be solved when passing to manifold control[27]. A possibility is to mimic natural systems by appropriately linking system degrees of freedom for better functioning, ie to make it fully "complex". Another class of problems is related to manipulation of information flux by itself in relation to the very fast development of systems handling this flux. Overall, a more sophisticated level is now appearing which relates to the shift toward more global properties "intelligent" systems should have. Recall that at each level of structure development, the system should satisfy specific properties represented in corresponding mathematical terms : stability for following its command, robustness for facing adverse environment, and at task level, determinism for guaranteeing that action is worth doing it. Prior to development of more powerful hardware components, the problem should be solved by proper embedding into the formalism of recent advances in modern functional analysis methods in order to evaluate the requirements for handling this new paradigm. An interesting output of proposed "functional" approach is to show that contrary to usual results, it is possible at lower system level to guarantee robust asymptotic stability within a robustness ball at least the size of system uncertainty. The method can even be extended to accommodate "unknown" systems for which there is reverse inequality[28]. The resulting two-level control is such that upper decisional level is totally freed from lower guiding (classical) level and can be devoted to manifold selection on which system trajectory takes place. Today active research is conducted on guaranteeing determinism by merging IT with complex systems with very difficult questions still under study[29] in concomitant manipulation of power and information fluxes, especially in embedded autonomous systems playing a larger and larger role in modern human civilisation. Summarizing, in many observed cases, systems from chemistry, physics, biology and probably economy to (artificial) man made ones are becoming dissipative as a consequence of a microscopic non accessible phenomenon which strongly determines global system behaviour. The resulting internal organization which adapts to boundary conditions from system environment leads to great difficulty for representing correctly their dynamics. For man made systems, the control of their dynamics is the more difficult as they ought to be more autonomous to comply with imposed constraints. Only a compromise is possible by acting on inputs respecting this internal organization.

4 Conclusion

The recent technology development has opened the exploration of a new kind of systems characterized by a large number of heterogeneous and strongly interacting components. These "complex" systems are exhibiting important enough self-organization to shield individual component behaviour in their response to exterior action which now takes place on specific manifolds corresponding to dynamic invariants generated by components interactions. The manifolds are the "smallest" element which can be acted upon from outside because even if finer elements can be observed, there is no information content associated to it for acting on the system. As there is a reduction of accessible system state space, the system at complex stage becomes more independent from outside environment power flux (and more robust to its variations). When associating information flux in primitive systems by creating "memory" DNA molecules very early in Earth history, Nature has extended system independence with respect to laws of Physics by existence of living beings (however the unavoidable dissipative nature of their structure leads to necessarily finite lifetime which Nature has been circumventing by inventing reproduction, transferring the advance in autonomy to species). In this sense, and in agreement with Aristoteles view, existence of complex systems is the first necessary step from natural background structure toward independence and isolation of a domain which could later manage its own evolution by accessing to life and finally to thought. Though at a minor level, a somewhat similar problem exists today with man made systems which have a large number of highly performing and strongly interacting components, and have to be optimized for economic reasons. To deal with such a situation, a control type approach has been chosen as most appropriate for properly setting the compromise between outer action and internal organization which now drives power fluxes inside the system. The problem of studying complex systems dynamics is expressible as an embedding problem in appropriate function spaces solvable by application of fixed point theorem, of which all previous results since Poincaré and Lyapounov are special applications. The control approach is also useful for mastering man made systems routinely built up today in industry and for giving them the appropriate structure implying today the delegation of specific "intelligence" required for guiding system trajectory. In parallel, the huge production increase and the multiplication of production centres is opening the questions of their interaction with environment and of the resulting risk, both domains in which scientific response has been modest up to now due to inadequacy of classical "hard" methods to integrate properly their global (and essential) aspect. Aside, the role of "soft" sciences has been also disappointing as long as they have not yet integrated quantitative observations in their approach.

References

[1] S. Johnson, *Emergence*, Penguin, New-York, 2001

M.A. Bedau, Weak Emergence, In J. Tomberlin (Ed.) *Philosophical Perspectives: Mind, Causation, and World*, Vol. 11, Blackwell pp.375-399, 1997

S. Strogatz, *Sync: The Emerging Science of Spontaneous Order*, 2003, Theia

G. Di Marzo Serugendo, *Engineering Emergent Behaviour: A Vision, Multi-Agent-Based Simulation III.* 4th International Workshop, MABS 2003 Melbourne, Australia, July 2003, David Hales et al. (Eds), LNAI 2927, Springer, 2003

E. Bonabeau, M. Dorigo, G. Theraulaz, *Swarm Intelligence: From Natural to Artificial Systems*, Oxford University Press, 1999, ISBN 0195131592

H. Nwana, *Software Agents : an Overview.* The Knowledge Engineering Review, Vol.11(3), pp.205-244, 1996

R. Conte, C. Castelfranchi, *Simulating multi agent interdependencies. A two-way approach to the micro-macro link*, in Klaus G. Troitzsch et al. (Eds.), Social science micro-simulation, Springer pp.394-415, 1995

S.A. DeLoach, *Multi-agent Systems Engineering: A Methodology And Language for Designing Agent Systems*, Procs. Agent-Oriented Information Systems '99 (AOIS'99), 1999

J. Ferber, O. Gutknecht, F. Michel, *From Agents To Organizations: An Organizational View of Multi- Agent Systems*, in "Agent-Oriented Software Engineering IV, AOSE 2003", Paolo Giorgini et al. (Eds.), LNCS 2935, Springer pp.214-230, 2004

C. Bernon, M.-P. Gleizes, S. Peyruqueou, G. Picard, ADELFE, *a Methodology for Adaptive Multi-Agent Systems Engineering*, 3rd International Workshop on "Engineering Societies in the Agents World III" (ESAW 2002), Petta, Tolksdorf & Zambonelli (Eds.) LNCS 2577, Springer pp.156-169, 2003

J. Fromm, *Types and Forms of Emergence*, http://arxiv.org/abs/nlin.AO/0506028

P. Bresciani, A. Perini, P. Giorgini, F. Giunchiglia, J. Mylopoulos, *A knowledge level software engineering methodology for agent oriented programming*, 5th International Conference on Autonomous Agents, ACM Press pp.648-655, 2001

J. Lind, *Iterative Software Engineering for Multi-agent Systems - The MASSIVE Method*, LNCS 1994, Springer, 2001

N. Jennings, *On Agent-Based Software Engineering*, Artificial Intelligence, Vol. 117(2) pp.277-296, 2000

J.O. Kephart, *Research Challenges of Autonomic Computing*, Proc. ICSE 05 pp.15-22, 2005

M. Kolp, P. Giorgini, J. Mylopoulos, *A goal-based organizational perspective on multi-agent architectures*, In Intelligent Agents VIII: Agent Theories,

Architectures, and Languages, LNAI 2333, Springer pp.128-140, 2002

P. Maes, *Modelling Adaptive Autonomous Agents*, http://citeseer.ist.psu.edu/42923.html

K.P. Sycara, *Multi-agent Systems*, AI Magazine Vol. 19 No. 2 (1998) 79-92

R. Abbott, *Emergence Explained*, to be published, see http//cs.calstatela.edu/ wiki/ images/9/90/Emergence-Explained.pdf

D. Yamins, *Towards a Theory of "Local to Global"* in Distributed Multi-Agent Systems, In Proceedings of the International Joint Conference on Autonomous Agents and Multi Agent Systems (AAMAS 2005). To appear

W. Vogels, R. van Renesse, K. Birman, *The Power of Epidemics: Robust Communication for Large-Scale Distributed Systems*, ACM SIGCOMM Computer Communication Review Vol. 33(1), pp.131-135, 2003

F. Zambonelli, N.R. Jennings, M. Wooldridge, *Developing Multi-agent Systems: The Gaia Methodology*, ACM Transactions on Software Engineering and Methodology, Vol. 12(3), pp.317-370, 2003

[2] M. Gell-Mann, *What is Complexity*, Complexity J., Vol.1(1), pp.16, 1995

The Quark and the Jaguar - Adventures in the simple and the complex, Little, Brown & Company, New-York, 1994

R.K. Standish, *On Complexity and Emergence*, Complexity International, Vol.9, pp.1-6, 2004

H. Morowitz, *The Emergence of Complexity*, ibid. pp.4

G. Nicolis, I. Prigogine, *A la Rencontre du Complexe*, PUF, Paris, 1992

J. Fromm, *The Emergence of Complexity*, Kassel University Press, 2004

S.Y. Auyang, *Foundations of Complex-system theories*, Cambridge University Press, 1998

J. Holland, *Hidden Order : How Adaptation Builds Complexity*, Helix Books, Addison Wesley, New-York, 1995

M. Waldrop, *Complexity : the Emerging Science at the Edge of Order and Chaos*, Touchstone, NY, 1993

Yu.L. Klimontovich, *Entropy, Information, and Criteria of Order in Open Systems*, Nonlinear Phenomena in Complex Systems, Vol.2(4), pp.1-25, 1999

L.O. Chua, *CNN : a Paradigm for Complexity*, World Scientific Publ., 1998

G. Nicolis, I. Prigogine, *Exploring Complexity : an Introduction*, W.H. Freeman and Co, NY, 1989

O. Babaoglu, M. Jelasity, A. Montresor et al. (Eds.), *Self-star Properties in Complex Information Systems: Conceptual and Practical Foundations*, Springer LNCS 3460, 2005

S.A. Kauffman, *The Origins of Order : Self-Organization and Selection in the Universe*, Oxford Univ. Press, 1993

R. Feistel, W. Ebeling, *Evolution of Complex Systems*, Kluver, Dordrecht, 1989

R. Badii and A. Politi, *Complexity: Hierarchical structures and scaling in physics*, Cambridge University Press, Cambridge, Mass., 1997

B. Goodwin, *How the Leopard Changed Its Spots: The Evolution of Complexity*, Weidenfield & Nicholson, London, 1994

B. Parker, *Chaos in the Cosmos : the Stunning Complexity of the Universe*, Plenum Press, 1996

G. Parisi, *Complex Systems : a Physicist's Viewpoint*, Internet arxiv:cond-mat/0205297

[3] W.R. Ashby, *Principles of the self-organizing system*, in "Principles of Self-Organization: Transactions of the University of Illinois Symposium", H. Von Foerster and G. W. Zopf, Jr. (eds.), Pergamon Press, pp. 255- 278, 1962

E. Jantsch, *Self Organizing Universe: Scientific and Human Implications*, Pergamon Press, 1980

A.D. Linkevich, *Self-organization in Intelligent Multi-agent Systems and Neural Networks*, Nonlinear Phenomena in Complex Systems, Part I : Vol.4(1), pp.18-46, 2001; Part II : Vol.4(3), pp.212-249, 2001

Per Bak, *How Nature Works: The Science of Self-Organized Criticality*, Copernicus Press, 1996

D.E.J. Blazis, *Introduction to the Proceedings of the Workshop "The Limits to Self-Organization in Biological Systems"*, The Biological Bulletin Vol. 202(3), pp.245-246, 2002

Jaegwon Kim, *Philosophy of Mind*, Westview Press, 1996

Wei-Min Shen et al., *Hormone-inspired self-organization and distributed control of robotic swarms*, Autonomous Robots, Vol.17, pp.93-105, 2004

W. Ebeling, *Entropy and Information in Processes of Self Organization : Uncertainty and Predictability*, Physica A, Vol.194(1-4), pp.563-575, 1993

S. Camazine, J.L. Deneubourg, N.R. Franks, J. Sneyd, G. Theraulaz, E. Bonabeau, *Self organization in Biological Systems*, Princeton Univ. Press, Princeton, New-Jersey, 2002

G. Nicolis, I. Prigogine, *Self Organization in Non Equilibrium Systems*, Wiley, New-York, 1977

G. Di Marzo Serugendo et al., *Engineering Self- Organising Systems*, LNAI 2977, Springer, 2004

Yu.L. Klimontovitch, *Criteria for Self Organization*, Chaos, Solitons & Fractals, Vol.5(10), pp.1985-1995, 2002

H. Haken, *Information and Self-Organization*, Springer, Berlin, 1988

F.E. Yates (ed) *Self-Organizing Systems: The Emergence of Order* , Plenum Press, 1987

S. Nolfi, D. Floreano, *Evolutionary Robotics : the Biology, Intelligence and Technology of Self-organizing Machines*, The MIT Press, Cambridge, Mass., 2000

R. Serra, M. Andretta, M. Compiani, G. Zanarini, *Introduction to the Physics of Complex Systems (the Mesoscopic Approach to Fluctuations, Nonlinearity and Self-Organization)*, Pergamon Press, 1986

[4] R.C. Hilborn, *Chaos and Nonlinear Dynamics*, Oxford Univ. Press, Oxford, UK, 1994

T. Kapitaniak, *Chaos for Engineers : Theory, Applications and Control*, Springer, Berlin, 1998

I. Prigogine, *Les Lois du Chaos*, Nouvelle Bibliothèque Scientifique, Flammarion, Paris, 1993

S.H. Strogatz, *Nonlinear Dynamics and Chaos*, Addison-Wesley, Reading, 1994

Y. Bar-Yam, *Dynamics of Complex Systems*, Addison-Wesley, Reading, 1997

[5] M. Hirsch, C. Pugh, M. Shub, *Invariant Manifolds*, Lecture Notes in Math. 583, Springer-Verlag, Berlin, 1977

A. Goriely, *Integrability and Nonintegrability of Dynamical Systems*, World Scientific Publ., 2001

[6] P.J. Antsaklis, K.M. Passino, *An Introduction to Intelligent and Autonomous Control*, Kluwer Acad. Publ., Norwell, MA, 1993

S. Russell, P. Norvig, *Artificial Intelligence : a Modern Approach*, Prentice-Hall, 1995

L.X. Wang, *Adaptive Fuzzy Systems and Control : Design and Stability Analysis*, Prentice-Hall, Englewood Cliffs, NJ, 1994

M. Brown, C. Harris, *Neuro Fuzzy Adaptive Modelling and Control*, Prentice Hall, Englewood Cliffs, NJ, 1994

B. Kosko, *Neural Networks and Fuzzy Systems: A Dynamical Systems Approach to Machine Intelligence.* Englewood Cliffs, NJ: Prentice-Hall, 1991

G. Weiss, *Multiagent Systems. A Modern Approach to Distributed Artificial Intelligence*, the MIT Press, Cambridge, Mass. 1999

E. Bonabeau, M. Dorigo, G. Theraulaz, *Swarm Intelligence : from Natural to Artificial Systems*, Oxford Univ. Press, New-York, 1999

[7] B. Kosko, *Global Stability of Generalized Additive Fuzzy Systems*, IEEE Trans. On Systems, Man, and Cybernetics, Part C, Vol.28(3), p.441, 1998

[8] M. Cotsaftis, *Recent Advances in Control of Complex Systems*, Survey Lecture, Proc. ESDA'96, Montpellier, France, ASME, Vol.I, (1996), p.1

[9] M. Cotsaftis, *Vision Limitation for Robot Deformation Control*, Proc. 5th Intern. Conf. on Mechatronics and Machine Vision in Practice (M2VIP), Nanjing, China, (1998), p.393

M. Cotsaftis, *Application of Energy Conservation to Control of Deformable Systems*, Proceedings 3rd Workshop on Systems Science and its Application, Beidaihe, China, (1998), p.42

[10] H.C. von Baeyer, *Maxwell's Demon*, Random House, 1998

[11] V.G. Majda, *Sobolev Spaces*, Springer-Verlag, New-York, 1985

L. Amerio, G. Prouse, *Almost-Periodic Functions and Functional Equations*, Van Nostrand-Reinhold, New-York, 1971

[12] E. Zeidler, *Nonlinear Functional Analysis and its Applications*, Vol.I, Springer-Verlag, New-York, 1986

D.R. Smart, *Fixed Point Theorems*, Cambridge Univ. Press, Mass., 1980

[13] A.M. Lyapounov, *Le Probleme General de la Stabilite du Mouvement*, Ann. Fac. Sciences Toulouse, 1907

[14] H. Poincare, *Les Methodes Nouvelles de la Mecanique Celeste*, 3 Vol., Gauthier-Villars, Paris, 1892-1899

[15] M. Cotsaftis, *Comportement et Controle des Systemes Complexes*, Diderot, Paris, 1997

[16] B.G. Pachpatte, *Mathematical Inequalities*, North Holland, Amsterdam, 2005

[17] M. Abramowitz, I.A. Stegun, *Handbook of Mathematical Functions*, Dover Publ., New-York, 1965

[18] M. Cotsaftis, *Popov Criterion Revisited for Other Nonlinear Systems*, Proc. ISIC 2003(International Symposium on Intelligent Control), Oct 5-8, Houston, 2003

[19] M. Cotsaftis, *Robust Asymptotic Control for Intelligent Unknown Mechatronic Systems*, ICARCV 2006, Singapore, Dec 5-8, 2006, to be published

[20] J. Appell, P.P. Zabrijko, *Nonlinear Superposition Operators*, Cambridge University Press, Mass., 1990.

[21] M. Farge, *Wavelet Transforms and their Applications to Turbulence*, Annu. Rev. Fluid Mech., Vol.24, pp.395-457,1992; M. Farge, K. Schneider, G. Pellegrino, A.A. Wray, R.S. Rogallo, Coherent Vortex Extraction in 3D Homogeneous Turbulence : Comparison between CVS-wavelet and POD-Fourier Decompositions, Phys. Fluids, Vol.15(10), pp.2886-2896

B. Nikolaenko, C. Foias, R.Temam, *The Connection between Infinite Dimensional and Finite Dimensional Dynamical Systems*, AMS, 1992

K.A. Morris, *Design of Finitedimensional Controllers for Infinite-dimensional Systems by Approximation*, J. of Mathematical Systems, Estimation and Control, Vol.4(2), pp.1-30, 1994

[22] D. Kondepudi, I. Prigogine, *Modern Thermodynamics : from Heat Engines to Dissipative Structures*, J. Wiley and Sons, NY, 1997

[23] S. Chandrasekhar, *Stochastic Problems in Physics and Astronomy*, Rev. Mod. Phys., Vol.15, no. 1, pp.1- 89, 1943

[24] M. Cotsaftis, *Lectures on Advanced Dynamics*, Taiwan Univ., 1993

[25] S. Lefschetz, *Stability of Nonlinear Control Systems*, Academic Press, NY, 1965

M.A. Aizerman, F.R. Gantmacher, *Absolute Stability of Regulator Systems*, Holden-Day, San Franscisco, 1964

G.E. Dullerud, F. Paganini, *A Course in Robust Control Theory : a Convex Approach*, Springer-Verlag, New-York, 2000

J.C.S. Hsu, A.U. Meyer, *Modern Control Principles and Applications*, McGraw-Hill, New-York, 1968

S. Arimoto, *Control Theory of Nonlinear Mechanical Systems : a Passivity Based and Circuit Theoretic Approach*, Oxford Univ. Press, Oxford, UK, 1996

J.H. Burl, *Linear Optimal Control, H_2 and H_∞ Methods*, Addison-Wesley Longman, Menlo Park, CA, 1999

G.A. Leonov, I.V. Ponomarenko, V.B. Smirnova, *Frequency Domain Methods for Nonlinear Analysis : Theory and Applications*, World Scientific Publ., Singapore, 1996

[26] M. Cotsaftis, *Beyond Mechatronics ; Toward Global Machine Intelligence*, Proc. ICMT¡¯2005, Kuala- Lumpur, Dec. 6-8, 2005

[27] M. Cotsaftis, *From Trajectory Control to Task Control . Emergence of Self Organization in Complex Systems*, in Emergent Properties in Natural and Artificial Dynamical Systems, M.A. Aziz-Alaoui, C. Bertelle, Eds, Springer-Verlag, pp.3-22, 2006

M. Cotsaftis, *Global Control of Flexural and Torsional Deformations of One-Link Mechanical System*, Kybernetika, Vol.33, no. 1, p.75, 1997

[28] M. Cotsaftis, *Robust Asymptotically Stable Control for Unknown Robotic Systems*, Proceedings 1998 Symposium on ISSPR, Hong Kong, Vol.I, p.267, 1998

[29] M. Cotsaftis, *Merging Information Technologies with Mechatronics . The Autonomous Intelligence Challenge*, Proc. IEECON¡¯2006, Nov.6-10, 2006, to be published

Holistic Metrics, a Trial on Interpreting Complex Systems

J. Manuel Feliz-Teixeira and António E. S. Carvalho Brito

Faculty of Engineering of University of Porto
Portugal
feliz@fe.up.pt, acbrito@fe.up.pt

Summary. In this text is proposed a simple method for estimating or characterize the behaviour of complex systems, in particular when these are being studied throughout simulation. Usual ways of treating the complex output data obtained from the activity (real or simulated) of such a kind of systems, which in many cases people classify and analyse along the time domain, usually the most complex perspective, is herein substituted by the idea of representing such data in the frequency domain, somehow like what is commonly done in Fourier Analysis and in Quantum Mechanics. This is expected to give the analyst a more holistic perspective on the system's behaviour, as well as letting him/her choose almost freely the complex states in which such behaviour is to be projected. We hope this will lead to simpler processes in characterizing complex systems.

1 Introduction

There are presently very few notes on the kind of metrics that could be reliable and of practical relevance when applied to the interpretation of complex systems behaviour. These systems are often based on intricate structures where a high number of entities interact with each other. Metrics are there for appropriately characterizing the nodes or individual parts of such structures, or small groups of them, but when the intent is a measure for the complete structure either they fail or appear to be too simplistic. That is certainly a good reason for modelling those cases using a strategic point of view, removing the time variable from the process, as in doing so the complexity is reduced a priori.

But when a dynamic and detailed representation is essential, the interpretation of the results and the characterization of the system frequently fail. This issue seems sometimes also related to a certain tendency impregnated in the minds to look at the systems from a pre-established perspective. At this point, however, perhaps this may be considered a conflict between different scientific approaches: the classical western reductionism, of anglo-saxonic inspiration, which believes the best approach is to break the system into small parts and understand, model or control those parts separately and then join them together, therefore looking at the world in an individualist way; and a more holistic approach, a vision slowly spreading and largely inspired by oriental cultures, which considers that each part of the system must be seen together with the whole and not in isolation, and therefore locates the tone in how the interactions between such parts contribute to the whole behaviour. Hopp and Spearman [Hopp et al. 2001], for instance, comment about this saying that "too much emphasis on individual components can lead to a loss of perspective for the overall system".

A significant number of authors defend this opinion, pointing out the importance of developing a more holistic point of view to interpret and study systems behaviour, in a way that analyses maintain enough fidelity to the system as a whole. As Tranouez et al. [Tranouez et al. 2003], who apply simulation to ecosystems, would say: a complex system is more than the simple collection of its elements. In management science, for instance, the "western" approach frequently generates difficulties at the interfaces between elements, typically of inventory or communication type. On the other hand, as just-in-time (JIT) systems give better emphasis to the relations and interactions and are continuously improving, the overall movements tend to be more harmonious. JIT already looks at systems in a certain holistic way. The same seems to be true in regard to other fields where simulation is applied, and mainly when the number of states to simulate is high.

2 Holistic measuring (a proposal)

But, what concerning metrics? How can one measure such a high number of states typically found in complex systems in order to effectively retrieve from them some sort of useful information?

As a metric is a characterization, we could think that maybe the modern Data Mining (DM) techniques could be extensively applied, for instance. These techniques use decision trees and other algorithms to discover hidden patterns in huge amounts of data, and are nowadays applied to almost any problem based on extensive data records, for instance, in e-Commerce for customer profile monitoring, in genetics research, in fraud detection, credit risk analysis, etc., and even for suspected "terrorist" detection (see Edelstein, [Edelstein 2001, Edelstein 2003]). However, they often imply the usage of high performance computers, sometimes with parallel processors, as well as huge computational resources to analyse GBytes or even TBytes of data. They are useful when any single record of data can be precious for the future result, and thus when all data must be analysed.

On the other hand, in many practical simulations a significant amount of data is not significant for the final conclusions, the simulation process is in itself a filter, and therefore such data may well be ignored in the outputs, even if it could have been essential to ensure the detailed simulation process to run. In the perspective of the author, maybe there is a way that could deserve some attention: the idea is to filter such data during the simulation execution and, at the same time, to turn the measures probabilistic by using an approach somehow inspired by the Fourier Analysis and the Quantum Mechanics. That is, to represent the overall system state (Ψ) in terms of certain base functions (Ψ_i), and then to measure the probabilities (α_i) associated with each of these functions. The interesting aspect of this is that each base state function (Ψ_i) could even be arbitrarily chosen by the analyst, and the probabilities (α_i) easily computed during the simulation process. Final results would then be summarised in some expression of the form:

$$\Psi = \alpha_1 \Psi_1 + \alpha_2 \Psi_2 + ... + \alpha_i \Psi_i + ... + \alpha_n \Psi_n \tag{1}$$

which could be interpreted as: there is a probability of α_1 that the system will be found in the state Ψ_1, a probability of α_2 that the system will be found in the state Ψ_2, etc. This would be the final measure of the system, in a sort of characterization of expectations under certain conditions. This also corresponds to projecting the system behaviour into the generalised vectors base of state functions (Ψ_i). The amounts α_i simply correspond to the values of those projections.

In Fourier Analysis, for instance, the complex behaviour observed in the time axis (see the example of figure 1) is substituted by the decomposition of such a signal into *sine* and *cosine* mathematical functions, and that way transferred to the frequency domain.

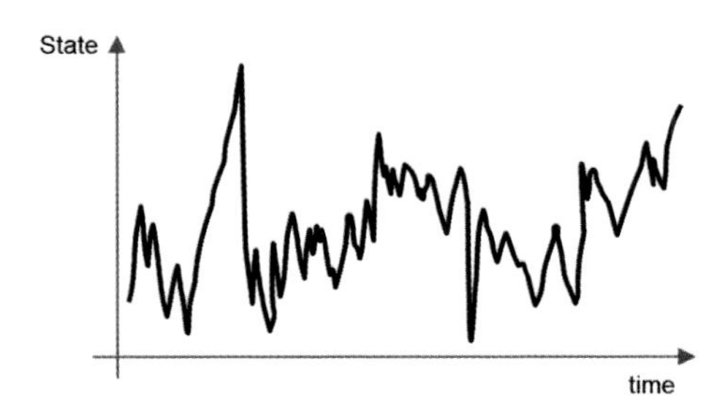

Fig. 1. Example of a general complex signal

The result is that the analyst is now much more able to visualize and to interpret the complexity of the previous signal, since it is as if this signal would be now expressed in terms of patterns (see example of figure 2). What firstly appeared as a confusing and almost randomly up-and-down behaviour may now be simply understood as the summation of some sinusoidal patterns with different amplitudes. Quantum Mechanics uses a similar formalism. We believe that the method proposed here will help generating such a clean view also when applied to the behaviour of complex systems.

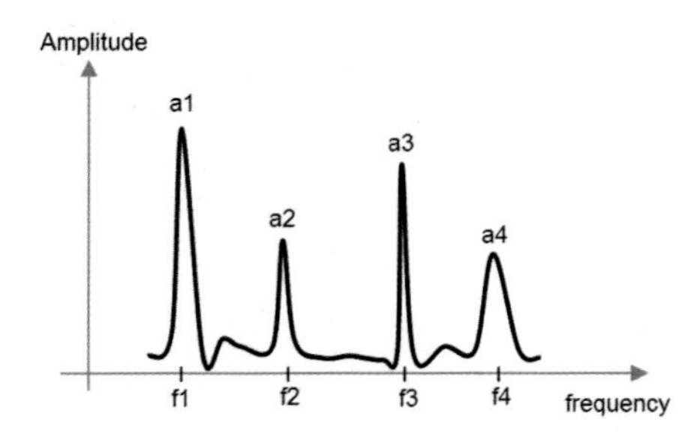

Fig. 2. Typical signal in the frequency domain

The present proposal may also be understood as an attempt to represent the system's behaviour in terms of a sort of generalised histogram, where the categories are the functions Ψ_i, which may correspond to the frequencies f_i in the previous figure, and the probabilities α_i are made to correspond to the amplitudes a_j in the same figure. In terms of this figure, the analyst would recognize a probability of a_1 that the system would be found in the state f_1, a probability of a_2 that the system would be found in the state f_2, etc.

3 An imaginary example

But, to help explain this, we can imagine a complex system like the Supply Chain shown in figure 3, for example. This is an example inspired by the company ZARA, the trendy Spanish clothes manufacturer of La Coruna. This company, from the INDITEX group, is worldwide known as a paradigm of success, despite its owner, and major manager, Mr Ortega, the second richest person in Spain, refusing several conventional practices claimed by most schools of management. ZARA refuses, for instance, the idea of advertisement. Forgive me if indirectly I am advertising it here. Returning to our subject, how can we apply our concept of holistic metrics to retrieve some useful information from such a complex case [1]? How can we specify the base functions

[1] In this figure is represented less than perhaps 10% of the real ZARA global Supply Chain structure.

(or base states) in which the system's behaviour will be projected? How will we calculate and represent the respective projections?

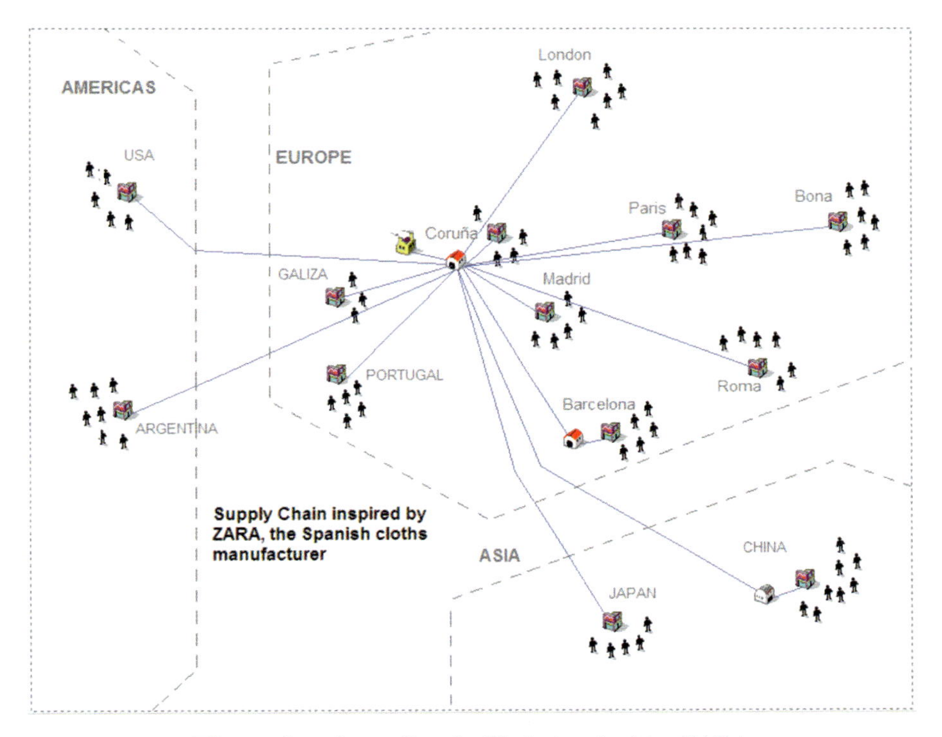

Fig. 3. Imaginary Supply Chain inspired by ZARA

First of all, we have to choose the Ψ_i functions into which the measures will be projected. We may choose them in terms of some specific *conditions* related to the information that must be obtained from the system. For example, if Mr Ortega is concerned about the levels of *stockouts*, *holdingcosts*, *servicelevel*, *turnover*, etc., which are typical measures of Supply Chain Management, he may for example define some sort of base functions by using conditions of the type:

Ψ_1 - *Stockouts* above 7%;
Ψ_2 - *Holdingcosts* above 5%;
Ψ_3 - *Servicelevel* under 75%;
Ψ_4 - *Turnover* under 2%.

Then, while the system is running, it must be *projected* into these set of functions, that is, the occurrences of each of these conditions must be counted up, whenever they are true. Thus, supposing n_j the accumulated number of occurrences of the condition Ψ_j, and N_j the total number of its samples, an estimation of α_j can simply be computed as:

$$\alpha_j = n_j/N_j \tag{2}$$

And the overall system state will therefore be expressed as:

$$\Psi = (n_1/N_1)\Psi_1 + (n_2/N_2)\Psi_2 + (n_3/N_3)\Psi_3 + (n_4/N_4)\Psi_4 \tag{3}$$

Notice that, in general, base functions are chosen to be orthogonal, or independent of each other, but in fact that is not a must for using this type of representation. One can also project a system into non orthogonal axis. As we said previously, such a measure may be seen as a characterization of expectations under certain conditions. The overall system state is, in reality, represented by the following weighted expression:

$$\begin{aligned} &\alpha_1 \times (Stockouts > 7) + \alpha_2 \times (Holdingcosts > 5) \\ &+\alpha_3 \times (Servicelevel < 75) + \alpha_4 \times (Turnover < 2) \end{aligned} \tag{4}$$

Now, if we build a histogram out of this data, we will characterize the system by means of a probabilistic graphical format, obtaining something of the type presented in the next figure (Fig.4), where the probabilities are the α_i.

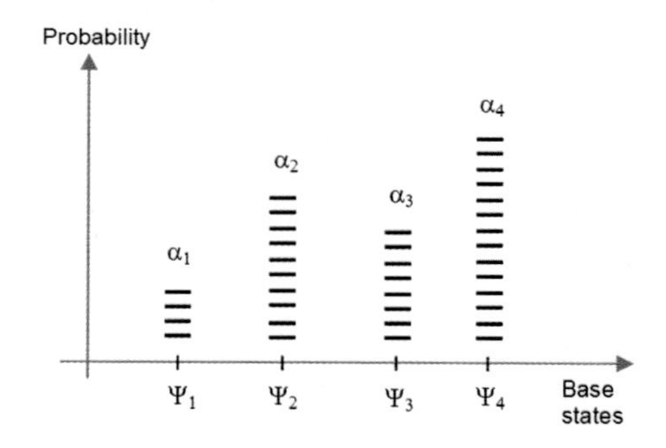

Fig. 4. Characterization of the system's behaviour

So, once the base states are well defined by the analyst, the characterization of the system is possible, no matters how complex the system is. We recall that in many practical cases the analyst is mainly focused in being sure that certain variables of the model do not cross some upper or lower limits, or, if they do, with which probability it happens. In order to evaluate the system in a wider range of modes of behaviour, several studies of this kind can be made with the system operating in different conditions. That will make possible to improve the knowledge about the system, or its characterization.

The former example was taken from a typical Supply Chain problem (see Feliz-Teixeira [Feliz-Teixeira 2006]), but this technique can be applied in general to other complex systems. For example, in a traffic system of a town, the complex states could be chosen to be the number of cars exceeding a certain value in a certain region, the travel time exceeding a certain value in another region, the number of public vehicles reaching a certain zone inferior to the minimum required, etc. As we recommend that these base functions (or complex base states) be well defined before simulation takes place, it implies that the simulation objectives must be well known prior to the start of the simulation process. Not always this is possible, of course, since simulation can be used to detect anomalous situations not predictable by means of other methods, for example.

This technique may, however, be also used as a method for analyse any sort of results, by being directly applied to the raw outputs of the complex system. In that case, the simulation will be a standard process and all the work is done by data manipulation. The results, in principle, will be the same, but that approach will in general be much more time consuming.

Finally, we would like to emphasise that we use the term "holistic metric" for distinguishing this kind of approach from those approaches which usually characterize systems by means of averages and standard deviations taken over a certain number of variables (usually a high number). These, as we know, frequently confuse the analyst's mind with the complexity of the results, instead of allowing a useful interpretation of the system's behaviour. Quantity of information is not all, and sometimes it can even generate confusion instead of clarity, if it is in excess. Besides, the method presented here goes on the trend of the "holistic" mind that seems to emerge in our days, as we defend.

4 Conclusions

Complex results generated by a complex system are very much dependent on how the analyst looks at the system and on how such results are analysed. We would say that any complex system can be minimally understood as long as the analyst knows what to search for, that is, if the objectives of the study are previously defined. This is because such objectives can in reality be used to establish the base functions (vectors) of an imaginary space where the complex behaviour will be projected, that way giving an automatic meaning to the results. This may also be seen as an attempt to measure the outputs of systems in the frequency domain (as in Fourier Analysis and in Quantum Mechanics), instead of in the time domain where signals usually are more difficult to interpret. Although no practical cases have yet been studied based on the idea presented in this article, we expect to use and test this approach in our next studies of simulation. We would also be pleased with receiving some feedback from anyone who decided to apply the same logic.

References

[Edelstein 2001] Edelstein, H. (2001) *Pan For Gold In The Clickstream.* InformationWeek.com.

[Edelstein 2003] Edelstein, H. (2003) *Using Data Mining to Find Terrorists.* Data Mining Review.

[Feliz-Teixeira 2006] Feliz-Teixeira, J. M. (2006) *Flexible Supply Chain Simulation (thesis).* Faculty of Engineering of Universitry of Porto, Portugal.

[Hopp et al. 2001] Hopp, W. J.; and M.L. Spearman (2001) *Factory Physics (second ed.).* Irwin McGraw-Hill.

[Tranouez et al. 2003] Tranouez, P.; Lerebourg, S.; Bertelle, C.; and D. Olivier (2003) *Changing the Level of Description in Ecosystem Models: an Overview.* 2003 European Simulation and Modelling Conference, Naples, Italy.

Different Goals in Multiscale Simulations and How to Reach Them

Pierrick Tranouez[1] and Antoine Dutot[2]

[1] MTG - UMR IDEES - Université de Rouen
IRED - 7 rue Thomas Becket
76821 Mont Saint Aignan Cedex - France
pierrick.tranouez@univ-rouen.fr
[2] LITIS - Université du Havre
25 rue Philippe Lebon - BP 540
76058 Le Havre Cedex - France
antoine.dutot@univ-lehavre.fr

Summary. In this paper we sum up our works on multiscale programs, mainly simulations. We first start with describing what multiscaling is about, how it helps perceiving signal from a background noise in a flow of data for example, for a direct perception by a user or for a further use by another program. We then give three examples of multiscale techniques we used in the past, maintaining a summary, using an environmental marker introducing an history in the data and finally using a knowledge on the behavior of the different scales to really handle them at the same time.

Key words: Multiscale, clustering, dynamic graphs, adaptation

1 Introduction: What this paper is about, and what it's not

Although we delved into different applications and application domains, the computer science research goals of our team has remained centered on the same subject for years. It can be expressed in different ways that we feel are, if not exactly equivalent, at least closely connected. It can be defined as managing multiple scales in a simulation. It also consists in handling emergent structures in a simulation. It can often also be seen as dynamic heuristic clustering of dynamic data[3]. This paper is about this theme, about why we think

[3] We will of course later on describe in more details what we mean by all this.

it is of interest and what we've done so far in this direction. It is therefore akin to a state of the art kind of article, except more centered on what we did. We will allude to what others have done, but the focus of the article is presenting our techniques and what we're trying to do, like most articles do, and not present an objective description of the whole field, as the different applications examples could make think : we're sticking to the same computer science principles overall. We're taking one step back from our works to contemplate them all, and not the three steps which would be necessary to encompass the whole domain, as it would take us beyond the scope of this book.

2 Perception: filtering to make decisions

I look at a fluid flow simulation but all I'm interested in is where does the turbulence happen, in a case where I couldn't know before the simulation [Tranouez 2005a]. I use a multi-participant communication system in a crisis management piece of software and I would like to know what are the main interests of each communicant based on what they are saying [Lesage 1999]. I use an Individual-Based Model (IBM) of different fish species but I'm interested in the evolution of the populations, not the individual fish [Prevost 2004]. I use a traffic simulation with thousands of cars and a detailed town but what I want to know is where the traffic jams are (coming soon).

In all those examples, I use a piece of software which produces huge amounts of data but I'm interested in phenomena of a different scale than the raw basic components. What we aim at is helping the user of the program to reach what he is interested in, be this user a human (Clarification of the representation) or another program (Automatic decision making). Although we're trying to stay general in this part, we focused on our past experience of what we actually managed to do, as described in "Some techniques to make these observations in a time scale comparable to the observed", this is not gratuitous philosophy.

2.1 Clarification of the representation

This first step of our work intends to extract the patterns on the carpet from its threads [Tranouez 1984]. Furthermore, we want it to be done in "real (program) time", meaning not a posteriori once the program is ended by examining its traces [Servat 1998], and sticking as close as possible to the under layer, the one pumping out dynamic basic data. We don't want the discovery of our structures to be of a greater time scale than a step of the program it works upon.

How to detect these structures? For each problem the structure must be analyzed, to understand what makes it stand out for the observer. This implies knowing the observer purpose, so as to characterize the structure. The answers are problem specific, nevertheless rules seem to appear.

In many situations, the structures are groups of more basic entities, which then leads to try to fathom what makes it a group, what is its inside, its outside, its frontier, and what makes them so.

Quite often in the situation we dealt with, the groups members share some common characteristics. The problem in that case belongs to a subgenre of clustering, where the data changes all the time and the clusters *evolve* with them, they are not computed from scratch at each change.

The other structures we managed to isolate are groups of strongly communicating entities in object-oriented programs like multiagent simulations. We then endeavored to manage these cliques.

In both cases, the detected structures are emphasized in the graphical representation of the program. This clarification lets the user of the simulation understand what happens in its midst. Because modeling, and therefore understanding, is clarifying and simplifying in a chosen direction a multi-sided problem or phenomenon, our change of representation participates to the understanding of the operator. It is therefore also a necessary part of automating the whole understanding, aiming for instance at computing an artificial decision making.

2.2 Automatic decision making

Just like the human user makes something of the emerging phenomena the course of the program made evident, other programs can use the detected organizations.

For example in the crisis management communication program, the detected favorite subject of interest of each of the communicant will be used as a filter for future incoming communications, favoring the ones on connected subjects. Other examples are developed below, but the point is once the structures are detected and clearly identified, the program can use models it may have of them to compute its future trajectory. It must be emphasized that at this point the structures can themselves combine into groups and structures of yet another scale, recursively. We're touching there an important component of complex system [Simon 1996]. We may hope the applications of this principle to be numerous, such as robotics, where perceiving structures in vast amounts of data relatively to a goal, and then being able to act upon these accordingly is a necessity.

We're now going to develop these notions in examples coming from our past works.

3 Some techniques to make these observations in a time scale comparable to the observed

The examples of handling dynamic organization we chose are taken from two main applications, one of a simulation of a fluid flow, the other of the simulation of a huge cluster of computed processes, distributed over a dynamic network of computing resources, such as computers. The methods titled "Maintaining a summary of a simulation" and "Reification: behavioral methods" are theories from the hydromechanics simulation, while "Traces of the past help understand the present" refers to the computing resources management simulation. We will first describe these two applications, so that an eventual misunderstanding of what they are does not hinder later the clarity of our real purpose, the analysis of multiscale handling methods.

In a part of a more general estuarine ecosystem simulation, we developed a simulation of a fluid flow. This flow uses a particle model [Leonard 1980], and is described in details in [Tranouez 2005a] or [Tranouez 2005b]. The basic idea is that each particle is a vorticity carrier, each interacting with all the others following Biot-Savart laws. As fluid flows tend to organize themselves in vortices, from all spatial scales from a tens of angstrom to the Atlantic Ocean, this is these vortices we tried to handle as the multiscale characteristic of our simulation. The two methods we used are described below.

The other application, described in depth in [Dutot 2005], is a step toward automatic distribution of computing over computing resources in difficult conditions, as:

- The resources we want to use can each appear and disappear, increase or decrease in number.
- The computing distributed is composed of different object-oriented entities, each probably a thread or a process, like in a multiagent system for example (the system was originally imagined for the ecosystem simulation alluded to above, and the entities would have been fish, plants, fluid vortices etc., each acting, moving ...)

Furthermore, we want the distribution to follow two guidelines:

- As much of the resources as possible must be used,
- Communications between the resources must be kept as low as possible, as it should be wished for example if the resources are computers and the communications therefore happen over a network, bandwidth limited if compared to the internal of a computer.

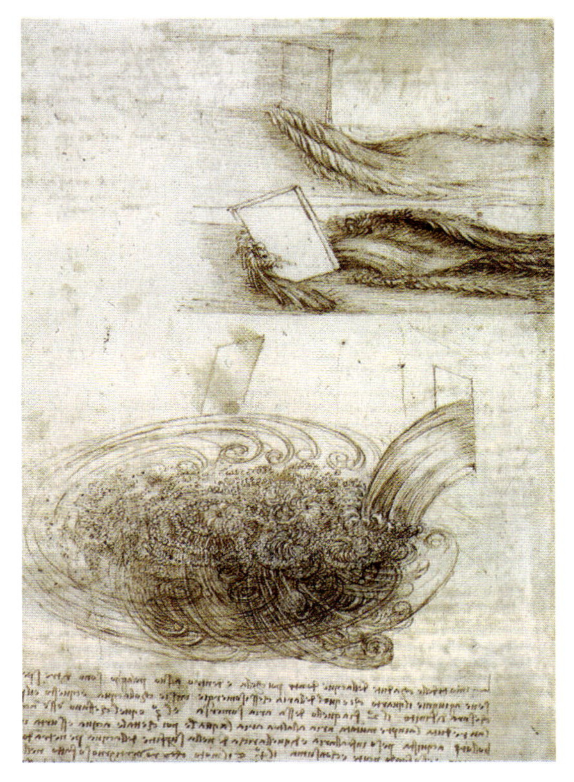

Fig. 1. Studies of water passing obstacles and falling by Leonardo Da Vinci, c. 1508-9. In Codex Leicester.

This the ultimate goal of this application, but the step we're interested in today consists in a simulation of our communicating processes, and of a program which, at the same time the simulated entities act and communicate, advises how they should be regrouped and to which computing resource they should be allocated, so as to satisfy the two guidelines above.

3.1 Maintaining a summary of a simulation

The first method we would like to describe here relates to the fluid flow simulation. The hydrodynamic model we use is based on an important number of interacting particles. Each of these influences all the others, which makes n^2 interactions, where n is the number of particles used. This makes a great number of computations. Luckily, the intensity of the influence is inversely proportional to the square of the distance separating two particles. We therefore use an approximation called Fast Multipoles Method (FMM), which consists in covering the simulation space with grids, of a density proportional to the density of particles (see Figure 2-a). The computation of the influence of its colleagues over a given particle is then done exactly for the ones close enough, and averaged on the grid for those further. All this is absolutely monoscale.

As the particles are vorticity carriers, it means that the more numerous they are in a region of space, the more agitated the fluid they represent is. We would therefore be interested in the structures built of close, dense particles, surrounded by sparser ones. A side effect of the grids of the FMM, is that they help us do just that. It is not that this clustering is much easier on the grids, it's above all that they are an order of magnitude less numerous, and organized in a tree, which makes the group detection much faster than if the algorithm was ran on the particles themselves. Furthermore, the step by step management of the grids is not only cheap (it changes the constant of the complexity of the particles movement method but not the order) but also needed for the FMM.

We therefore detect structures on

- Dynamic data (the particles characteristics)
- With little computing added to the simulation,

which is what we aimed at.

The principle here is that through the grids we maintain a summary of the simulation, upon which we can then run static data algorithm, all this at a cheap computing price.

3.2 Traces of the past help understand the present

The second method relates to the detection of communication clusters inside a distributed application. The applications we are interested in are composed of a large number of object-oriented entities that execute in parallel, appear, evolve and, sometimes, disappear. Aside some very regular applications, often entities tend to communicate more with some than with others. For example in a simulation of an aquatic ecosystem, entities representing a species of fish may stay together, interacting with one another, but flee predators. Indeed organizations appear groups of entities form. Such simulations are a good example of applications we intend to handle, where the number of entities is often too large to compute a result in an acceptable time on one unique computer.

To distribute these applications it would be interesting to both have approximately the same number of entities on each computing resource to balance the load, but also to avoid as much as possible to use the network, that costs significantly more in terms of latency than the internals of a computer. Our goal is therefore to balance the load and minimize network communications.

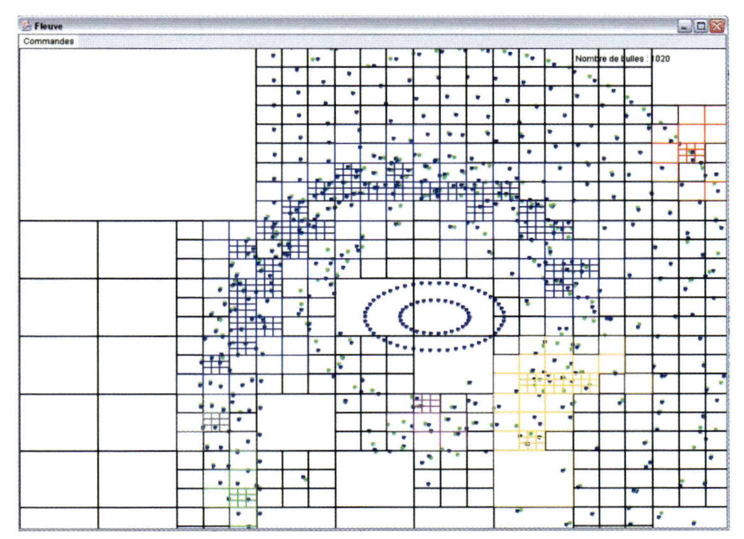

a - Each color corresponds to a detected aggregate

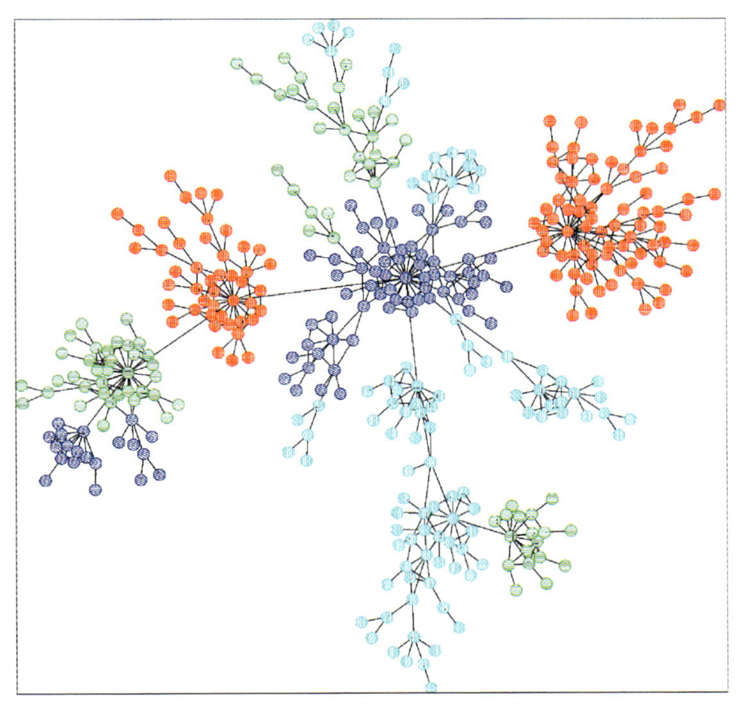

b - Each color corresponds to a computing ressource

Fig. 2. Detection of emergent structures in two applications with distinct methods

Sadly, these criteria are conflicting, and we must find a tradeoff.

Our method consists in the use of an ant metaphor. Applications we use are easily seen as a graph, which is a set of connected entities. We can map entities to vertices of the graph, and communications between these entities to the edges of the graph. This graph will follow the evolution of the simulation. When an entity appear, a vertex will appear in the graph, when a communication will be established between two entities, an edge will appear between the two corresponding vertices. We will use such a graph to represent the application, and will try to find clusters of highly communicating entities in this graph by coloring it, assigning a color to each cluster. This will allow to identify clusters as a whole and use this information to assign not entities, but at another scale, clusters to computing resources (Figure 2-b).

For this, we use numerical ants that crawl the graph as well as their pheromones, olfactory messages they drop, to mark clusters of entities. We use several distinct colonies of ants, each of a distinct color, that drop colored pheromones. Each color corresponds to one of the computing resources at our disposal. Ants drop colored pheromones on edges of the graph when they cross them. We mark a vertex as being of the color of the dominant pheromone on each of its incident edges. The color indicates the computing resource where the entity should run.

To ensure our ants color groups of highly communicating entities of the same color to minimize communications, we use the collaboration between ants: ants are attracted by pheromones of their own color, and attracted by highly communicating edges. To ensure the load is balanced, that is to ensure that the whole graph is not colored only in one color if ten colors are available, we use competition, ants are repulsed by the pheromones of other colors.

Pheromones in nature being olfactory molecules, they tend to evaporate. Ants must maintain them so they do not disappear. Consequently, only the interesting areas, zones where ants are attracted, are covered by pheromones and maintained. When a zone becomes less interesting, ants leave it and pheromone disappear. When an area becomes of a great interest, ants colonize it by laying down pheromones that attract more ants, and the process self-amplifies.

We respect the metaphor here since it brings us the very interesting property of handling the dynamics. Indeed, our application continuously changes, the graph that represents it follows this, and we want our method to be able to discover new highly communicating clusters, while abandoning vertices that are no more part of a cluster. As ants continuously crawl through the graph, they maintain the pheromone color on the highly communicating clusters. If entities and communications of the simulation appear or disappear, ants can quickly adapt to the changes. Colored pheromones on parts where a cluster disappeared evaporate and ants colonize new clusters in a dynamic way. Indeed, the application never changes completely all the time; it modifies itself smoothly. Ants lay down "traces" of pheromones and do not recompute the color of each vertex at each time, they reuse the already dropped pheromone therefore continuously giving a distribution advice at a small computing price, and adapting to the reconfigurations of the underlying application.

3.3 Reification: behavioral methods

This last example of our multiscale handling methods was also developed on the fluid flow simulation (Figure 3). Once more, we want to detect structures in a dynamic flow of data, without getting rid of the dynamicity by doing a full computation on each step of the simulation. The idea here is doing the full computation only once in a good while, and only relatively to the unknown parts of our simulation.

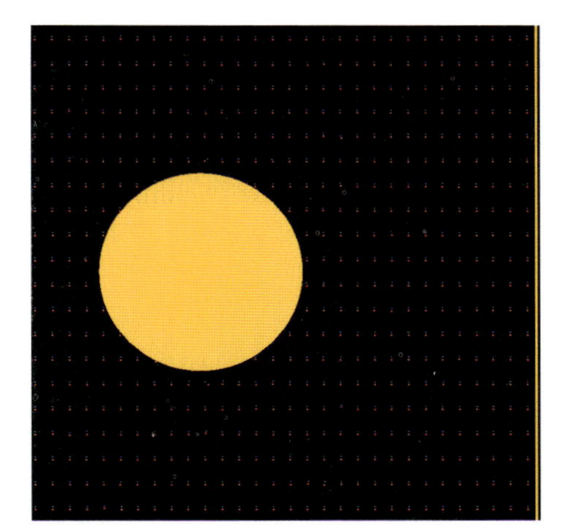

Fig. 3. Fluid flow around an obstacle. On the left, the initial state. On the right, a part of the flow, some steps later (the ellipses are vortices)

We begin with detecting vortices on the basic particles once. Vortices will be a rather elliptic set of close particles of the same rotation sense. We then introduce a multiagent system of the vortices (Figure 3-right). We have indeed a general knowledge of the way vortices behave. We know they move like a big particle in our Biot-Savard model, and we model its structural stability through social interactions with the surrounding basic particles, the other vortices and the obstacles, through which they can grow, shrink or die (be dissipated into particles). The details on this can be found in [Tranouez 2005a]. Later on, we occasionally make a full-blown vortex detection, but only on the remaining basic particles, as the already detected vortexes are managed by the multiagent system.

In this case, we possess knowledge on the structures we want to detect, and we use it to build actually the upper scale level of the simulation, which at the same time lightens ulterior structures detection. We are definitely in the category described in Automatic decision making.

4 Conclusion

Our research group works on complex systems and focuses on the computer representation of their hierarchical/holarchical characteristics [Koestler 1978], [Simon 1996], [Kay 2000]. We tried to illustrate that describing a problem at different scales is a well-spread practice at least in the modeling and simulating community. We then presented some methods for handling the different scales, with maintaining a summary, using an environmental marker introducing a history in the data and finally using knowledge on the behavior of the different scales to handle them at the same time.

We now believe we start to have sound multiscale methods, and must focus on the realism of the applications, to compare the sacrifice in details we make when we model the upper levels rather than just heavily computing the lower ones. We save time and lose precision, but what is this trade-off worth *precisely*?

References

[Dutot 2005] Dutot, A. (2005) *Distribution dynamique adaptative à l'aide de mécanismes d'intelligence collective.* PhD thesis, Le Havre University.

[Kay 2000] Kay, J. (2000) *Ecosystems as Self-Organising Holarchic Open Systems : narratives and the second law of thermodynamics.* Jorgensen, S.E.; and F. Müller (Eds.), Handbook of Ecosystems Theories and Management, Lewis Publishers.

[Koestler 1978] Koestler, A. (1978) *Janus. A Summing Up.* Vintage Books, New York.

[Leonard 1980] Leonard, A. (1980) *Vortex methods for flow simulation*, Journal of Computational Physics, vol. 37, 289-335.

[Lesage 1999] Lesage, F.; Cardon, A.; and P. Tranouez (1999) *A multiagent based prediction of the evolution of knowledge with multiple points of view*, KAW'99.

[Prevost 2004] Prevost, G.; Tranouez, P.; Lerebourg, S.; Bertelle, C. and D. Olivier (2004) *Methodology for holarchic ecosystem model based on ontological tool.* ESM 2004, 164-171.

[Servat 1998] Servat, D.; Perrier, E.; Treuil, J.-P.; and A. Drogoul (1998) *When Agents Emerge from Agents: Introducing Multi-scale Viewpoints in Multi-agent Simulations.* MABS 98, 183-198.

[Simon 1996] Simon, H. (1996) *The Sciences of the Artificial (3rd Edition).* MIT Press.

[Tranouez 1984] Tranouez, Pierre (1984) *Fascination et narration dans l'œuvre romanesque de Barbey d'Aurevilly.* Doctorat d'État.

[Tranouez 2005a] Tranouez, P.; Bertelle, C; and D. Olivier (2006) *Changing levels of description in a fluid flow simulation* in M.A. Aziz-Alaoui and C. Bertelle (eds), "Emergent Properties in Natural and Artificial Dynamical Systems", Understanding Complex Systems series, 87-99.

[Tranouez 2005b] Tranouez, P. (2005) *Contribution à la modélisation et à la prise en compte informatique de niveaux de descriptions multiples. Application aux écosystèmes aquatiques (Penicillo haere, nam scalas aufero)*, PhD thesis, Le Havre University.

Invariant Manifolds of Complex Systems

Jean-Marc Ginoux and Bruno Rosseto

PROTEE Laboratory
IUT de Toulon, Université du Sud, BP 20132
83957 La Garde cedex, France
ginoux@univ-tln.fr, rossetto@univ-tln.fr

Summary. The aim of this work is to establish the existence of invariant manifolds in *complex systems*. Considering *trajectory curves* integral of multiple time scales dynamical systems of dimension two and three (predator-prey models, neuronal bursting models) it is shown that there exists in the phase space a *curve* (resp. a *surface*) which is invariant with respect to the flow of such systems. These invariant manifolds are playing a very important role in the stability of complex systems in the sense that they are "restoring" the determinism of *trajectory curves*.

Key words: Invariant curves, invariant surfaces, multiple time scales dynamical systems, complex systems.

1 Dynamical systems

In the following we consider a system of ordinary differential equations defined in a compact E included in $\mathbb{R}^n$

$$\frac{d\mathbf{X}}{dt} = \Im(\mathbf{X}) \tag{1}$$

with $\mathbf{X} = [x_1, x_2, ..., x_n]^t \in E \subset \mathbb{R}^n$ and

$$\Im(\mathbf{X}) = [f_1(\mathbf{X}), f_2(\mathbf{X}), ...f_n(\mathbf{X})]^t \in E \subset \mathbb{R}^n .$$

The vector $\Im(\mathbf{X})$ defines a velocity vector field in E whose components f_1 which are supposed to be continuous and infinitely differentiable with respect to all x_t and t, i.e., are C^∞ functions in E and with values included in $\mathbb{R}$. For more details, see for example [Coddington et al. 1955]. A solution of this system is an *integral curve* $\mathbf{X}(t)$ tangent to $\Im$ whose values define the states of the dynamical system described by Equation (1). Since none of the components f_i of the velocity vector field depends here explicitly on time, the system is said to be autonomous.

2 Trajectory curves

The integral of the system (1) can be associated with the coordinates, i.e., with the position, of a point M at the instant t. The total derivative of $\mathbf{V}(t)$ namely the instantaneous acceleration vector $\boldsymbol{\gamma}(t)$ may be written, while using the chain rule, as:

$$\gamma = \frac{\mathrm{d}\mathbf{V}}{\mathrm{d}t} = \frac{\mathrm{d}\Im}{\mathrm{d}\mathbf{X}}\frac{\mathrm{d}\mathbf{X}}{\mathrm{d}t} = J\mathbf{V} \tag{2}$$

where $\frac{\mathrm{d}\Im}{\mathrm{d}\mathbf{X}}$ is the functional jacobian matrix J of the system (1). Then, the *integral curve* defined by the vector function $\mathbf{X}(t)$ of the scalar variable t representing the trajectory of M can be considered as a *plane* or a *space curve* which has local metric properties namely *curvature* and *torsion*. A n-dimensional trajectory curve has $(n-1)$ curvatures. For $n = 2$, the first curvature is called curvature while for $n = 3$ the second curvature is called torsion. In what follows $\|.\|$ represents the Euclidean norm.

2.1 Curvature

The curvature, which expresses the rate of changes of the tangent to the trajectory curve of system (1), is defined by

$$\frac{1}{\Re} = \frac{\|\gamma \wedge \mathbf{V}\|}{\|\mathbf{V}\|^3} \tag{3}$$

where $\Re$ represents the *radius of curvature.*

2.2 Torsion

The *torsion*, which expresses the difference between the *trajectory curve* of system (1) and a *plane curve*, is defined by:

$$\frac{1}{\Im} = -\frac{\dot{\gamma} \cdot (\gamma \wedge \mathbf{V})}{\|\gamma \wedge \mathbf{V}\|^2} \tag{4}$$

where $\Im$ represents the *radius of torsion.*

3 Lie Derivative — Darboux Invariant

Let φ a C^4 function defined in a compact E included in $\mathbb{R}$ and $\mathbf{X}(t)$ the integral of the dynamic system defined by (1). The Lie derivative is defined as follows:

$$L_{\mathbf{X}}\varphi = \mathbf{V} \cdot \boldsymbol{\nabla}\varphi = \sum_{i=1}^{n} \frac{\partial \varphi}{\partial x_i}\dot{x}_i = \frac{\mathrm{d}\varphi}{\mathrm{d}t} \tag{5}$$

Theorem 1. *An invariant curve (resp. surface) is defined by* $\varphi(\mathbf{X}) = 0$ *where* φ *is a* C^1 *in an open set* U *and such there exists a* C^4 *function denoted* $k(\mathbf{X})$ *and called cofactor which satisfies*

$$L_{\mathbf{X}}\phi(\mathbf{X}) = k(\mathbf{X})\phi(\mathbf{X}) \tag{6}$$

for all $\mathbf{X} \in U$.

Proof of this theorem may be found in [Darboux 1878].

Theorem 2. *If* $L_{\mathbf{X}}\varphi = 0$ *then* φ *is first integral of the dynamical system defined by (1). So,* φ *is first integral of the dynamical system defined by* $\{\varphi = \alpha\}$ *and where* α *is constant.*

Proof of this theorem may be found in [Demazure 1989].

4 Invariant Manifolds

According to the previous theorems 1 and 2 the following proposition may be established.

Proposition 1. *The location of the points where the local curvature of the trajectory curves integral of a two dimensional dynamical system defined by (1) vanishes is first integral of this system. Moreover, the invariant curve thus defined is over flowing invariant with respect to the dynamical system (1).*

Proof of this theorem may be found in [Ginoux et al. 2006].

Proposition 2. *The location of the points where the local torsion of the trajectory curves integral of a three dimensional dynamical system defined by (1) vanishes is first integral of this system. Moreover, the invariant surface thus defined is over flowing invariant with respect to the dynamical system (1).*

Proof of this theorem may be found in [Ginoux et al. 2006].

5 Applications to Complex Systems

According to this method it is possible to show that any dynamical system defined by (1) possess an invariant manifold which is endowing stability with the trajectory curves, restoring thus the loss determinism inherent to the non-integrability feature of these systems. So, this method may be also applied to any complex system such that predator-prey models, neuronal bursting models... But, in order to give the most simple and consistent application, let's focus on two classical examples:

- the Balthazar Van der Pol model;
- the Lorenz model.

5.1 Van der Pol model

The oscillator of B. Van der Pol [Van der Pol 1926] is a second-order system with non-linear frictions which can be written:

$$\ddot{x} + \alpha\left(x^2 - 1\right)\dot{x} + x = 0\,.$$

The particular form of the friction which can be carried out by an electric circuit causes a decrease of the amplitude of the great oscillations and an increase of the small. There are various manners of writing the previous equation like a first order system. One of them is:

$$\begin{cases} \dot{x} = \alpha\left(x + y - \dfrac{x^3}{3}\right) \\ \dot{y} = -\dfrac{x}{\alpha} \end{cases}$$

When α becomes very large, x becomes a fast variable and y a slow variable. In order to analyze the limit $\alpha \to \infty$, we introduce a small parameter $\epsilon = \frac{1}{\alpha^2}$ and a slow time $t' = \frac{t}{\alpha}\sqrt{\epsilon t}$. Thus, the system can be written:

$$\mathbf{V} = \begin{pmatrix} \dfrac{\mathrm{d}x}{\mathrm{d}t} \\ \dfrac{\mathrm{d}y}{\mathrm{d}t} \end{pmatrix} = \Im \begin{pmatrix} f(x,y) \\ g(x,y) \end{pmatrix} = \begin{pmatrix} \dfrac{1}{\epsilon}\left(x + y - \dfrac{x^3}{3}\right) \\ -x \end{pmatrix} \tag{7}$$

with ϵ a positive real parameter: $\epsilon = 0.05$ and where the functions f and g are infinitely differentiable with respect to all x_i and t, i.e. are C^∞ functions in a compact E included in $\mathbb{R}^2$ and with values in $\mathbb{R}$.
According to Proposition 1, the location of the points where the local *curvature* vanishes leads to the following equation:

$$\phi(x,y) = 9y^2 + \left(9x + 3x^3\right)y + 6x^4 - 2x^6 + 9x^2\epsilon \tag{8}$$

According to Theorem 1 (Cf. Appendix for details), the Lie derivative of Equation (8) may be written:

$$L_{\mathbf{X}}\phi(\mathbf{X}) = \mathrm{Tr}[J]\phi(\mathbf{X}) + \frac{2x^2}{\epsilon}\left(-3x - 3y + x^3\right) \tag{9}$$

Let's plot the function $\phi(x,y)$ (in blue), its Lie derivative $L_{\mathbf{X}}\phi(\mathbf{X})$ (in magenta), the *singular approximation* $x + y - \frac{x^3}{3}$ (in green) and the *limit cycle* corresponding to system (7) (in red):
According to Fenichel's theory, there exists a function $\varphi(x,y)$ defining a manifold (curve) which is overflowing invariant and which is $C^r\mathcal{O}(\epsilon)$ close to the *singular approximation*. It is easy to check that in the vicinity of the *singular approximation* which corresponds to the second term of the right-hand-side of Equation (9) we have:

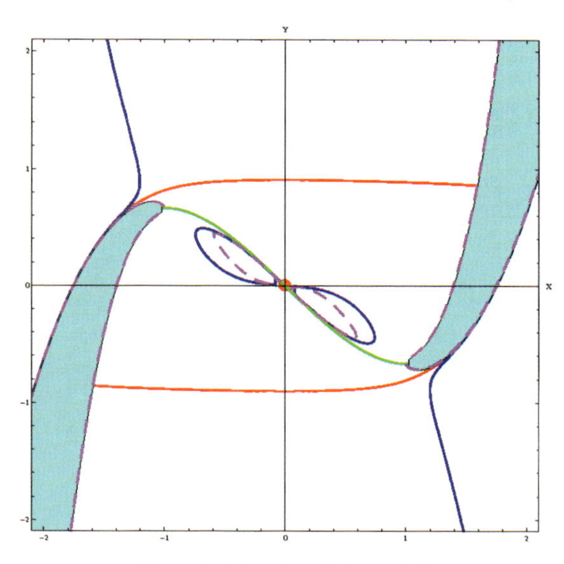

Fig. 1. Van der Pol model.

$$L_{\mathbf{X}}\phi(\mathbf{X}) = \mathrm{Tr}[J]\phi(\mathbf{X})\,.$$

Moreover, it can be shown that in the location of the points where the local *curvature* vanishes, i.e., where $\varphi(x, y) = 0$. Equation (9) can be written:

$$L_{\mathbf{X}}\phi(\mathbf{X}) = 0\,.$$

So, according to Theorem 1 and 2, we can claim that the manifold defined by $\varphi(x, y) = 0$ is an *invariant curve* with respect to the flow of system (7) and is a *local first integral* of this system.

5.2 Lorenz model

The purpose of the model established by Edward Lorenz [Lorenz 1963] was in the beginning to analyze the impredictible behaviour of weather. It most widespread form is as follows:

$$\mathbf{V} = \begin{pmatrix} \dfrac{\mathrm{d}x}{\mathrm{d}t} \\ \dfrac{\mathrm{d}y}{\mathrm{d}t} \\ \dfrac{\mathrm{d}z}{\mathrm{d}t} \end{pmatrix} = \Im \begin{pmatrix} f(x,y,z) \\ g(x,y,z) \\ h(x,y,z) \end{pmatrix} = \begin{pmatrix} \sigma(y-x) \\ -xz + rx - y \\ xy - \beta z \end{pmatrix} \tag{10}$$

with σ, r and β are real parameters: $\sigma = 10$, $\beta = \frac{8}{3}$, $r = 28$ and where the functions f, g and h are infinitely differentiable with respect to all x_i, and t, i.e., are C^∞ functions in a compact E included in $\mathbb{R}^3$ and with values in $\mathbb{R}$. According to Proposition 1, the location of the points where the local torsion vanishes leads to an equation which for place reasons can not be expressed. Let's name it as previously:

$$\varphi(x, y, z)\,. \tag{11}$$

According to Theorem 1 (Cf. Appendix for details), the Lie derivative of Equation (11) may be written:

$$L_{\mathbf{X}}\phi(\mathbf{X}) = \mathrm{Tr}[J]\phi(\mathbf{X}) + P(\mathbf{V} \cdot \boldsymbol{\gamma}) \tag{12}$$

where P is a polynomial function of both vectors $\mathbf{V}$ and $\boldsymbol{\gamma}$. Let's plot the function $\phi(x, y, z)$ and its Lie derivative $L_{\mathbf{X}}\phi(\mathbf{X})$ and the *attractor* corresponding to system (10):

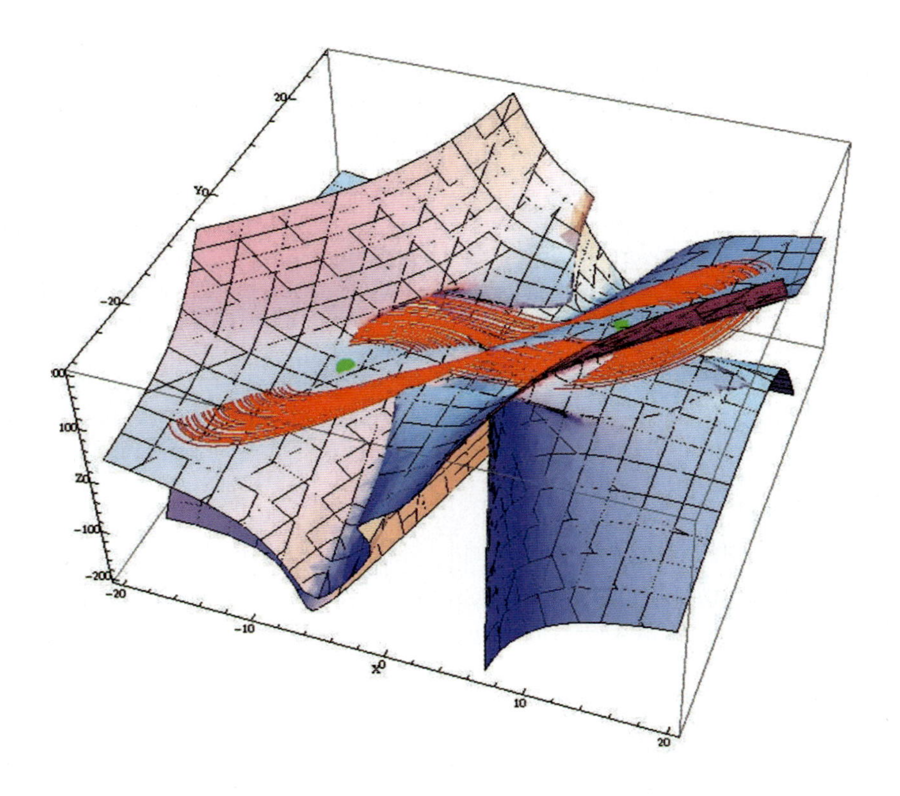

Fig. 2. Lorenz model.

It is obvious that the function $\phi(x, y, z)$ defining a manifold (surface) is merged into the corresponding to its Lie derivative. It is easy to check that in the vicinity of the manifold $\phi(x,y,z)$ Equation (12) reduces to:

$$L_{\mathbf{X}}\phi(\mathbf{X}) = \mathrm{Tr}[J]\phi(\mathbf{X}) .$$

Moreover, it can be shown that in the location of the points where the local torsion vanishes, i.e., where $\phi(x,y,z) = 0$ Equation (12) can be written:

$$L_{\mathbf{X}}\phi(\mathbf{X}) = 0$$

So, according to Theorem 1 and 2, we can claim that the manifold defined by $\phi(x,y,z) = 0$ is an invariant surface with respect to the flow of system (10) and is a local first integral of this system.

6 Discussion

In this work, existence of *invariant manifolds* which represent local first integrals of two (resp. three) dimensional dynamical systems defined by (1) has been established. From these two characteristics it can be stated that the former implies that such manifolds are representing the stable part of the trajectory curves in the phase space and from the latter that they are restoring the loss determinism inherent to the non-integrability feature of such systems. Moreover, while considering that dynamical systems defined by (1) include complex systems, it is possible to apply this method to various models of ecology (predator-prey models), neuroscience (neuronal bursting models), molecular biology (enzyme kinetics models)... Research of such invariant manifolds in coupled systems or in systems of higher dimension (four and more) would be of great interest.

Acknowledgements

Authors would like to thank Professors M. Aziz-Alaoui and C. Bertelle for their useful collaboration.

References

[Coddington et al. 1955] Coddington, E.A.; and N. Levinson (1955) *Theory of Ordinary Differential Equations.* Mac Graw Hill, New York.

[Darboux 1878] Darboux, G. (1878) *Mémoire sur les équations différentielles algébriques du premier ordre et du premier degré.* Bull. Sci. Math. Sér. **2** (2), 60-96, 123-143, 151-200.

[Demazure 1989] Demazure, M. (1989) *Catastrophes et Bifurcations.* Ellipses, Paris.

[Fenichel 1979] Fenichel, N. (1979) *Geometric singular perturbation theory for ordinary differential equations.* J. Diff. Eq., **31**, 53-98.

[Ginoux et al. 2006] Ginoux, J.M.; and B. Rossetto (2006) *Differential Geometry and Mechanics: applications to chaotic dynamical systems.* International Journal of Bifurcation and Chaos, **16(4)**, 887-910.
[Lorenz 1963] Lorenz, E. N. (1963) *Deterministic non-periodic flows.* J. Atmos. Sc., **20**, 130-141.
[Van der Pol 1926] Van der Pol, B. (1926) *On 'Relaxation-Oscillations'.* Phil. Mag., **7**, Vol. 2, 978-992.

Appendix

First of all, let's recall the following results:

$$L_{\mathbf{X}} \|\mathbf{u}\| = \frac{\mathrm{d}\|\mathbf{u}\|}{\mathrm{d}t} = \frac{\mathbf{u} \cdot \dot{\mathbf{u}}}{\|\mathbf{u}\|} \tag{13}$$

Two-dimensional dynamical system

Let's pose: $\varphi(\mathbf{X}) = \|\boldsymbol{\gamma} \wedge \mathbf{V}\|$. According to (13) the Lie derivative of this expression may be written:

$$L_{\mathbf{X}} \varphi(\mathbf{X}) = \frac{\mathrm{d}\|\boldsymbol{\gamma} \wedge \mathbf{V}\|}{\mathrm{d}t} = \frac{(\boldsymbol{\gamma} \wedge \mathbf{V}) \cdot \frac{\mathrm{d}}{\mathrm{d}t}(\boldsymbol{\gamma} \wedge \mathbf{V})}{\|\boldsymbol{\gamma} \wedge \mathbf{V}\|} \tag{14}$$

where $\frac{\mathrm{d}}{\mathrm{d}t}(\boldsymbol{\gamma} \wedge \mathbf{V}) = \dot{\boldsymbol{\gamma}} \wedge \mathbf{V}$.

According to Equation (2) the Lie derivative of the acceleration vector may be written:

$$\dot{\boldsymbol{\gamma}} = J\boldsymbol{\gamma} + \frac{\mathrm{d}J}{\mathrm{d}t}\mathbf{V} \tag{15}$$

it leads to:

$$\begin{aligned} \frac{\mathrm{d}}{\mathrm{d}t}(\boldsymbol{\gamma} \wedge \mathbf{V}) = \dot{\boldsymbol{\gamma}} \wedge \mathbf{V} &= \left(J\boldsymbol{\gamma} + \frac{\mathrm{d}J}{\mathrm{d}t}\mathbf{V} \right) \wedge \mathbf{V} \\ &= J\boldsymbol{\gamma} \wedge \mathbf{V} + \tfrac{\mathrm{d}J}{\mathrm{d}t}\mathbf{V} \wedge \mathbf{V} \end{aligned} \tag{16}$$

Using the following identity:

$$(J\mathbf{a}) \wedge \mathbf{b} + \mathbf{a} \wedge (J\mathbf{b}) = \mathrm{Tr}(J)(\mathbf{a} \wedge \mathbf{b})$$

it can be established that:

$$J\boldsymbol{\gamma} \wedge \mathbf{V} = \mathrm{Tr}(J)(\boldsymbol{\gamma} \wedge \mathbf{V})$$

So, expression (14) may be written:

$$L_{\mathbf{X}}\varphi(\mathbf{X}) = \frac{1}{\|\boldsymbol{\gamma} \wedge \mathbf{V}\|} \left(\mathrm{Tr}(J) \left(\boldsymbol{\gamma} \wedge \mathbf{V}\right) \cdot \left(\boldsymbol{\gamma} \wedge \mathbf{V}\right) + \left(\frac{\mathrm{d}J}{\mathrm{d}t}\mathbf{V} \wedge \mathbf{V}\right) \cdot \left(\boldsymbol{\gamma} \wedge \mathbf{V}\right) \right) \tag{17}$$

Let's note that: $(\boldsymbol{\gamma} \wedge \mathbf{V}) \cdot (\boldsymbol{\gamma} \wedge \mathbf{V}) = \|\boldsymbol{\gamma} \wedge \mathbf{V}\|^2$ and that: $\boldsymbol{\beta} = \frac{\boldsymbol{\gamma} \wedge \mathbf{V}}{\|\boldsymbol{\gamma} \wedge \mathbf{V}\|}$. So, equation (17) leads to:

$$L_{\mathbf{X}}\varphi(\mathbf{X}) = \mathrm{Tr}(J)\|\boldsymbol{\gamma} \wedge \mathbf{V}\| + \left(\frac{\mathrm{d}J}{\mathrm{d}t}\mathbf{V} \wedge \mathbf{V}\right) \cdot \boldsymbol{\beta} \tag{18}$$

Since vector $\frac{\mathrm{d}J}{\mathrm{d}t}\mathbf{V} \wedge \mathbf{V}$ has a unique coordinate according to the $\boldsymbol{\beta}$-direction and since we have posed: $\varphi(\mathbf{X}) = \|\boldsymbol{\gamma} \wedge \mathbf{V}\|$, expression (18) may finally be written:

$$L_{\mathbf{X}}\varphi(\mathbf{X}) = \mathrm{Tr}(J)\varphi(\mathbf{X}) + \left\|\frac{\mathrm{d}J}{\mathrm{d}t}\mathbf{V} \wedge \mathbf{V}\right\| \tag{19}$$

Three-dimensional dynamical system

Let's pose: $\varphi(\mathbf{X}) = \dot{\boldsymbol{\gamma}} \cdot (\boldsymbol{\gamma} \wedge \mathbf{V})$. The Lie derivative of this expression may be written:

$$L_{\mathbf{X}}\varphi(\mathbf{X}) = \frac{\mathrm{d}\left[\dot{\gamma} \cdot (\boldsymbol{\gamma} \wedge \mathbf{V})\right]}{\mathrm{d}t} \tag{20}$$

According to $\frac{\mathrm{d}}{\mathrm{d}t}\left[\dot{\gamma} \cdot (\boldsymbol{\gamma} \wedge \mathbf{V})\right] = \ddot{\boldsymbol{\gamma}} \cdot (\boldsymbol{\gamma} \wedge \mathbf{V})$, it leads to:

$$L_{\mathbf{X}}\varphi(\mathbf{X}) = \frac{\mathrm{d}\left[\dot{\gamma} \cdot (\boldsymbol{\gamma} \wedge \mathbf{V})\right]}{\mathrm{d}t} = \ddot{\boldsymbol{\gamma}} \cdot (\boldsymbol{\gamma} \wedge \mathbf{V}) \tag{21}$$

The Lie derivative of expression (15) leads to:

$$\ddot{\boldsymbol{\gamma}} = J\dot{\boldsymbol{\gamma}} + 2\frac{\mathrm{d}J}{\mathrm{d}t}\boldsymbol{\gamma} + \frac{\mathrm{d}^2 J}{\mathrm{d}t^2}\mathbf{V}$$

Thus, expression (21) reads:

$$L_{\mathbf{X}}\varphi(\mathbf{X}) = (J\dot{\boldsymbol{\gamma}}) \cdot (\boldsymbol{\gamma} \wedge \mathbf{V}) + \left(2\frac{\mathrm{d}J}{\mathrm{d}t}\boldsymbol{\gamma} + \frac{\mathrm{d}^2 J}{\mathrm{d}t^2}\mathbf{V}\right) \cdot (\boldsymbol{\gamma} \wedge \mathbf{V}) \tag{22}$$

It can also be established that:

$$\left(J^2\boldsymbol{\gamma}\right) \cdot (\boldsymbol{\gamma} \wedge \mathbf{V}) = \mathrm{Tr}(J)\left(J\boldsymbol{\gamma}\right) \cdot (\boldsymbol{\gamma} \wedge \mathbf{V})$$

So, since we have posed: $\varphi(\mathbf{X}) = \dot{\boldsymbol{\gamma}} \cdot (\boldsymbol{\gamma} \wedge \mathbf{V})$, expression (22) may finally be written:

$$L_{\mathbf{X}}\varphi(\mathbf{X}) = \mathrm{Tr}(J)\varphi(\mathbf{X}) + \left(-\mathrm{Tr}(J)\frac{\mathrm{d}J}{\mathrm{d}t}\mathbf{V} + J\frac{\mathrm{d}J}{\mathrm{d}t}\mathbf{V} + 2\frac{\mathrm{d}J}{\mathrm{d}t}\boldsymbol{\gamma} + \frac{\mathrm{d}^2 J}{\mathrm{d}t^2}\mathbf{V}\right) \cdot (\boldsymbol{\gamma} \wedge \mathbf{V}) \tag{23}$$

Application of Homotopy Perturbation Method for Ecosystems Modelling

Zaid Odibat[1] and Cyrille Bertelle[2]

[1] Prince Abdullah Bin Ghazi Faculty of Science and IT
Al-Balqa'Applied University
Salt, Jordan
odibat@bau.edu.jo

[2] LITIS, University of Le Havre
25 rue Philippe Lebon, BP 540
76058 Le Havre cedex, France
cyrille.bertelle@univ-lehavre.fr

Summary. The HPM method can be considered as one of the new methods belonging to the general classification of perturbation methods. These methods deals with exact solvers for linear differential equations and approximative solvers for non linear equations. In this paper, we focus our attention on the generation of the decomposition steps to build a solver using the HPM method. We present how this method can be used in ecosystem modelling. We develop some solvers for prey-predator systems involving 2 or 3 populations.

Key words: Homotopy perturbation method; ecosystems; Adomian decomposition method; variational iteration method.

1 Introduction

Ecosystems modelling can be approach in many ways. Global methods are based on differential systems and Individual-based methods allow to represent local phenomena. The first methods are efficient to formulate general behavior of the whole system by using global parameters. The second deal with a better understanding on which local phenomena are hidden inside these global parameters but they lead to high consuming computing. Innovative models are today based on hybrid approaches which manage during computing different way to express the ecosystems behavior. When and where some regularities are observed, we can change from individual-based models to differential ones. When automatic processes are developped in that way, we need both to build automatic equation and identify the parameters value but we need also to solve automatically the generated equations. The method proposed here is the Homotopy Perturbation Method (HPM) which is applied to nonlinear ecosystems.

Usually, perturbation methods need some kind of small parameter to be used. In the HPM method, which doesn't require a small parameter in an equation, a homotopy with an imbedding parameter $p \in [0,1]$ is constructed. The method provides analytical approximate solutions for different types of nonlinear ecosystems. The results reveal that the method is very effective and simple for obtaining approximate solutions of nonlinear systems of differential equations.

The HPM, proposed first by He [He 1999, He 2000], for solving differential and integral equations, linear and nonlinear, has been the subject of extensive analytical and numerical studies. The method, which is a coupling of the traditional perturbation method and homotopy in topology, deforms continuously to a simple problem which is easily solved. This method has a significant advantage in that it provides an analytical approximate solution to a wide range of nonlinear problems in applied sciences. The HPM is applied to Volterra's integro-differential equation [El-Shahed 2005], to nonlinear oscillators [He 2004a], bifurcation of nonlinear problems [He 2005a], bifurcation of delay-differential equations [He 2005b], nonlinear wave equations [He 2005c], boundary value problems [He 2006a], quadratic Riccati differential equation of fractional order [Odibat et al. 2008], and to other fields. This HPM yields a very rapid convergence of the solution series in most cases, usually only a few iterations leading to very accurate solutions.

2 Analysis of HPM

The HPM which provides an analytical approximate solution is applied to various nonlinear problems (see the references). In this section, we introduce a reliable algorithm to handle in a realistic and efficient way the nonlinear ecosystems. The proposed algorithm will then be used to investigate the system,

$$
\begin{aligned}
Dx_1(t) &= \textstyle\sum_{j=1}^{n} a_{1j}(t)x_j + f_1(t, x_1, x_2, \ldots, x_n), \\
Dx_2(t) &= \textstyle\sum_{j=1}^{n} a_{2j}(t)x_j + f_2(t, x_1, x_2, \ldots, x_n), \\
&\vdots \\
Dx_n(t) &= \textstyle\sum_{j=1}^{n} a_{nj}(t)x_j + f_n(t, x_1, x_2, \ldots, x_n),
\end{aligned}
\tag{1}
$$

subject to the initial condition,

$$x_1(0) = c_1, \quad x_2(0) = c_2, \ \ldots, x_n(0) = c_n, \tag{2}$$

where f_i is a nonlinear function for $i = 1, 2, \ldots, n$. In view of the homotopy perturbation technique, we can construct, for $i = 1, 2, \ldots, n$, the following homotopy,

$$Dx_i(t) - \sum_{j=1}^{n} a_{ij}(t)x_j = pf_i(t, x_1, x_2, \ldots, x_n), \tag{3}$$

where $p \in [0, 1]$. The homotopy parameter p always changes from zero to unity. In case of $p = 0$, Eq. (3) becomes the linear equation,

$$Dx_i(t) = \sum_{j=1}^{n} a_{ij}(t)x_j, \tag{4}$$

and when it is one, Eq. (3) turns out to be the original equation given in the system (1). The basic assumption is that the solution of the system (1) can be written as a power series in p:

$$x_i = x_i^{(0)} + px_i^{(1)} + p^2x_i^{(2)} + p^3x_i^{(3)} + \ldots \quad . \tag{5}$$

Substituting Eq. (5) into Eq. (3), and equating the terms with identical powers of p, we can obtain a series of linear equations of the form

$$
\begin{aligned}
&p^0 : Dx_i^{(0)}(t) = \textstyle\sum_{j=1}^{n} a_{ij}(t)x_j^{(0)},\ x_i^{(0)}(0) = c_i, \\
&p^1 : Dx_i^{(1)}(t) = \textstyle\sum_{j=1}^{n} a_{ij}(t)x_j^{(1)} + f_i^{(1)}(t, x^{(0)}),\ x_i^{(1)}(0) = 0, \\
&p^2 : Dx_i^{(2)}(t) = \textstyle\sum_{j=1}^{n} a_{ij}(t)x_j^{(2)} + f_i^{(2)}(t, x^{(0)}, x^{(1)}),\ x_i^{(2)}(0) = 0, \\
&p^3 : Dx_i^{(3)}(t) = \textstyle\sum_{j=1}^{n} a_{ij}(t)x_j^{(3)} + f_i^{(3)}(t, x^{(0)}, x^{(1)}, x^{(2)}),\ x_i^{(3)}(0) = 0, \\
&\quad \vdots
\end{aligned}
$$

where the functions $f_i^{(1)}, f_i^{(2)}, f_i^{(3)}, \ldots$ satisfy the following equation

$$f_i(t, x_1^{(0)} + px_1^{(1)} + p^2x_1^{(2)} + \ldots, \ldots, x_n^{(0)} + px_n^{(1)} + p^2x_n^{(2)} + \ldots)$$

$$= f_i^{(1)}(t, x_1^{(0)}, x_2^{(0)}, \ldots, x_n^{(0)}) + pf_i^{(2)}(t, x_1^{(0)}, x_2^{(0)}, \ldots, x_n^{(0)}, x_1^{(1)}, x_2^{(1)}, \ldots, x_n^{(1)})$$

$$+p^2 f_i^{(3)}(t, x_1^{(0)}, x_2^{(0)}, \ldots, x_n^{(0)}, x_1^{(1)}, x_2^{(1)}, \ldots, x_n^{(1)}, x_1^{(2)}, x_2^{(2)}, \ldots, x_n^{(2)}) + \ldots.$$

Setting $p = 1$ in the Eq. (5) yields the solution of the system (1). It is obvious that the above linear equations are easy to solve, and the components $x_i^{(k)}, k \geq 0$ of the homotopy perturbation solution can be completely determined, thus the series solution is entirely determined.
Finally, we approximate the solution $x_i(t) = \sum_{k=0}^{\infty} p^k x_i^{(k)}(t)$ by the truncated series

$$\phi_i(t) = \sum_{k=0}^{N-1} p^k x_i^{(k)}(t). \tag{6}$$

It is also useful, for the system (1), to construct the homotopy, for $i = 1, 2, \ldots, n$,

$$Dx_i(t) - p\sum_{j=1}^{n} a_{ij}(t)x_j = pf_i(t, x_1, x_2, \ldots, x_n), \tag{7}$$

where $p \in [0, 1]$. In this case, the term $\sum_{j=1}^{n} a_{ij}(t)x_j^{(0)}$ is combined with the component $x_i^{(1)}$ and the term $\sum_{j=1}^{n} a_{ij}(t)x_j^{(1)}$ is combined with the component $x_i^{(2)}$ and so on. This variation reduces the number of terms in each component and may minimize the size of calculations. Substituting (5) into (7), we obtain the following series of linear equations

$$p^0 : Dx_i^{(0)}(t) = 0,\ x_i^{(0)}(0) = c_i,$$

$$p^1 : Dx_i^{(1)}(t) = \textstyle\sum_{j=1}^{n} a_{ij}(t)x_j^{(0)} + f_i^{(1)}(t, x^{(0)}),\ x_i^{(1)}(0) = 0,$$

$$p^2 : Dx_i^{(2)}(t) = \textstyle\sum_{j=1}^{n} a_{ij}(t)x_j^{(1)} + f_i^{(2)}(t, x^{(0)}, x^{(1)}),\ x_i^{(2)}(0) = 0,$$

$$p^3 : Dx_i^{(3)}(t) = \textstyle\sum_{j=1}^{n} a_{ij}(t)x_j^{(2)} + f_i^{(3)}(t, x^{(0)}, x^{(1)}, x^{(2)}),\ x_i^{(3)}(0) = 0.$$

$$\vdots$$

3 Numerical Implementation

To demonstrate the effectiveness of the HPM algorithm discussed above, several examples of nonlinear systems will be studied. In the first example we choose a linear system to show the features of HPM and the convergence of the homotopy perturbation solution.

Example 1 *Consider the linear system [Momani et al. 2008]*

$$\begin{aligned} x'(t) &= y(t), \\ y'(t) &= 2x(t) - y(t), \end{aligned} \tag{8}$$

subject to the initial conditions

$$x(0) = 1 \quad , \quad y(0) = -1. \tag{9}$$

According to the homotopy given in Eq. (7), Substituting (5) and the initial conditions (9) into the homotopy (7) and equating the terms with identical powers of p, we obtain the following two sets of linear equations:

$$\begin{aligned} p^0: &\quad Dx^{(0)} = 0, && x^{(0)}(0) = 1, \\ p^1: &\quad Dx^{(1)} = y^{(0)}, && x^{(1)}(0) = 0, \\ p^2: &\quad Dx^{(2)} = y^{(1)}, && x^{(2)}(0) = 0, \\ p^3: &\quad Dx^{(3)} = y^{(2)}, && x^{(3)}(0) = 0, \\ &\quad \vdots \end{aligned}$$

$$\begin{aligned} p^0: &\quad Dy^{(0)} = 0, && y^{(0)}(0) = -1, \\ p^1: &\quad Dy^{(1)} = 2x^{(0)} - y^{(0)}, && y^{(1)}(0) = 0, \\ p^2: &\quad Dy^{(2)} = 2x^{(1)} - y^{(1)}, && y^{(2)}(0) = 0, \\ p^3: &\quad Dy^{(3)} = 2x^{(2)} - y^{(2)}, && y^{(3)}(0) = 0, \\ &\quad \vdots \end{aligned}$$

Consequently, solving the above equations, the first few components of the homotopy perturbation solution for the system (8) are derived as follows

$$
\begin{aligned}
x^{(0)} &= 1 \quad , \quad y^{(0)} = -1, \\
x^{(1)} &= -t \quad , \quad y^{(1)} = 3t, \\
x^{(2)} &= \tfrac{3}{2}t^2 \quad , \quad y^{(2)} = t - \tfrac{5}{2}t^2, \\
x^{(3)} &= -\tfrac{5}{6}t^3 \quad , \quad y^{(3)} = \tfrac{11}{6}t^3, \\
x^{(4)} &= \tfrac{11}{24}t^4 \quad , \quad y^{(4)} = -\tfrac{21}{24}t^4, \\
x^{(5)} &= -\tfrac{21}{120}t^5 \quad , \quad y^{(5)} = \tfrac{43}{120}t^5, \\
&\vdots
\end{aligned}
$$

and so on, in this manner the rest of components of the homotopy perturbation solution for the system (8) can be obtained. The solution in series form is given by

$$
\begin{aligned}
x(t) &= 1 - t + \tfrac{3}{2}t^2 - \tfrac{5}{6}t^3 + \tfrac{11}{24}t^4 - \tfrac{21}{120}t^5 + \ldots, \\
&= \tfrac{2}{3}\left(1 - 2t + \tfrac{(-2t)^2}{2!} + \tfrac{(-2t)^3}{3!} + \tfrac{(-2t)^4}{4!} + \tfrac{(-2t)^5}{5!} + \ldots\right) \\
&+ \tfrac{1}{3}\left(1 + t + \tfrac{t^2}{2!} + \tfrac{t^3}{3!} + \tfrac{t^4}{4!} + \tfrac{t^5}{5!} + \ldots\right),
\end{aligned}
\tag{10}
$$

$$
\begin{aligned}
y(t) &= -1 + 3t - \tfrac{5}{2}t^2 + \tfrac{11}{6}t^3 - \tfrac{21}{24}t^4 + \tfrac{43}{120}t^5 + \ldots, \\
&= -\tfrac{4}{3}\left(1 - 2t + \tfrac{(-2t)^2}{2!} + \tfrac{(-2t)^3}{3!} + \tfrac{(-2t)^4}{4!} + \ldots\right) \\
&+ \tfrac{1}{3}\left(1 + t + \tfrac{t^2}{2!} + \tfrac{t^3}{3!} + \tfrac{t^4}{4!} + \tfrac{t^5}{5!} + \ldots\right),
\end{aligned}
\tag{11}
$$

which converges to the exact solution

$$
\begin{aligned}
x(t) &= \tfrac{2}{3}e^{-2t} + \tfrac{1}{3}e^{t}, \\
y(t) &= -\tfrac{4}{3}e^{-2t} + \tfrac{1}{3}e^{t}.
\end{aligned}
\tag{12}
$$

Example 2 *Consider the predator-prey system [Momani et al. 2008]*

$$
\begin{aligned}
Dx(t) &= x(t) - x(t)y(t), \\
Dy(t) &= -y(t) + x(t)y(t),
\end{aligned}
\tag{13}
$$

subject to the initial conditions

$$
x(0) = 1 \quad , \quad y(0) = 0.5. \tag{14}
$$

According to the homotopy given in Eq. (7), Substituting (5) and the initial conditions (14) into the homotopy (7) and equating the terms with identical powers of p, we obtain the following two sets of linear equations:

$$\begin{aligned}
p^0 &: \quad Dx^{(0)} = 0, \qquad x^{(0)}(0) = 1,\\
p^1 &: \quad Dx^{(1)} = x^{(0)} - x^{(0)}y^{(0)}, \qquad x^{(1)}(0) = 0,\\
p^2 &: \quad Dx^{(2)} = x^{(1)} - x^{(0)}y^{(1)} - x^{(1)}y^{(0)}, \qquad x^{(2)}(0) = 0,\\
p^3 &: \quad Dx^{(3)} = x^{(2)} - x^{(0)}y^{(2)} - x^{(1)}y^{(1)} - x^{(2)}y^{(0)}, \qquad x^{(3)}(0) = 0,\\
&\quad \vdots
\end{aligned}$$

$$\begin{aligned}
p^0 &: \quad Dy^{(0)} = 0, \qquad y^{(0)}(0) = 0.5,\\
p^1 &: \quad Dy^{(1)} = -y^{(0)} + x^{(0)}y^{(0)}, \qquad y^{(1)}(0) = 0,\\
p^2 &: \quad Dy^{(2)} = -y^{(1)} + x^{(0)}y^{(1)} + x^{(1)}y^{(0)}, \qquad y^{(2)}(0) = 0,\\
p^3 &: \quad Dy^{(3)} = -y^{(2)} + x^{(0)}y^{(2)} + x^{(1)}y^{(1)} + x^{(2)}y^{(0)}, \qquad y^{3}(0) = 0,\\
&\quad \vdots
\end{aligned}$$

Consequently, solving the above equations, the first few components of the homotopy perturbation solution for the system (13) are derived as follows

$$\begin{aligned}
x^{(0)} &= 1, \quad , \quad y^{(0)} = \tfrac{1}{2},\\
x^{(1)} &= \tfrac{1}{2}t \quad , \quad y^{(1)} = 0,\\
x^{(2)} &= \tfrac{1}{8}t^2 \quad , \quad y^{(2)} = \tfrac{1}{8}t^2,\\
x^{(3)} &= -\tfrac{1}{48}t^3 \quad , \quad y^{(3)} = \tfrac{1}{48}t^3,\\
&\vdots
\end{aligned}$$

and so on, in this manner the rest of components of the homotopy perturbation solution can be obtained. The fourth-term approximate solution for the system (13) is given by

$$\begin{aligned}
x(t) &= 1 + \tfrac{1}{2}t + \tfrac{1}{8}t^2 - \tfrac{1}{48}t^3,\\
y(t) &= \tfrac{1}{2} + \tfrac{1}{8}t^2 + \tfrac{1}{48}t^3,
\end{aligned} \tag{15}$$

which is the same solution for the system (13) obtained in [Momani et al. 2008] using Adomian decomposition method and variational iteration method.

Example 3 *Consider the predator-prey system*

$$\begin{aligned} Dx(t) &= ax(t) - bx(t)y(t), \\ Dy(t) &= -cy(t) + dx(t)y(t), \end{aligned} \tag{16}$$

where a, b, c and d are constants, subject to the initial conditions

$$x(0) = c_1 \quad , \quad y(0) = c_2. \tag{17}$$

This system is a generalization of the system (13). Using the homotopy given in Eq. (3), the third-term approximate solution for the system (16) is given by

$$\begin{aligned} x(t) = {} & c_1 \exp(at) + \tfrac{bc_1c_2}{c}\Big(\exp((a-c)t) - \exp(at)\Big) \\ & -\tfrac{bdc_1^2c_2}{a}\Big(\tfrac{\exp((2a-c)t)}{a-c} + \tfrac{\exp((a-c)t)}{c}\Big) \\ & -\tfrac{b^2c_1c_2^2}{c}\Big(-\tfrac{\exp((a-2c)t)}{2c} + \tfrac{\exp((a-c)t)}{c}\Big) \\ & +\Big(\tfrac{bdc_1^2c_2}{a}\Big[\tfrac{1}{a-c} + \tfrac{1}{c}\Big] + \tfrac{b^2c_1c_2^2}{2c^2}\Big) \exp(at), \\ y(t) = {} & c_2 \exp(-ct) + \tfrac{dc_1c_2}{a}\Big(\exp((a-c)t) - \exp(-ct)\Big) \\ & +\tfrac{d^2c_1^2c_2}{a}\Big(\tfrac{\exp((2a-c)t)}{2a} - \tfrac{\exp((a-c)t)}{a}\Big) \\ & +\tfrac{bdc_1c_2^2}{c}\Big(\tfrac{\exp((a-2c)t)}{a-c} - \tfrac{\exp((a-c)t)}{a}\Big) \\ & +\Big(\tfrac{d^2c_1^2c_2}{2a^2} - \tfrac{bdc_1c_2^2}{c}\Big[\tfrac{1}{a-c} - \tfrac{1}{a}\Big]\Big) \exp(-ct). \end{aligned} \tag{18}$$

Example 4 *Consider the predator-prey system*

$$\begin{aligned} Dx(t) &= ax(t) - bx(t)y(t) - cx(t)z(t), \\ Dy(t) &= -dy(t) + ex(t)y(t) - fy(t)z(t), \\ Dz(t) &= -gz(t) + hx(t)z(t) + iy(t)z(t), \end{aligned} \tag{19}$$

where a, b, c, d, e, f, g, h and i are constants, subject to the initial conditions

$$x(0) = c_1 \quad , \quad y(0) = c_2 \quad , \quad z(0) = c_3. \tag{20}$$

According to the homotopy given in Eq. (3), Substituting (5) and the initial conditions (20) into the homotopy (3) and equating the terms with identical powers of p, we obtain the following two sets of linear equations:

$$\begin{aligned} p^0 : \quad & Dx^{(0)} = ax^{(0)}, \qquad x^{(0)}(0) = c_1, \\ p^1 : \quad & Dx^{(1)} = ax^{(1)} - bx^{(0)}y^{(0)} - cx^{(0)}z^{(0)}, \qquad x^{(1)}(0) = 0, \\ p^2 : \quad & Dx^{(2)} = ax^{(2)} - b(x^{(0)}y^{(1)} + x^{(1)}y^{(0)}) - c(x^{(0)}z^{(1)} + x^{(1)}z^{(0)}), \\ & \vdots \end{aligned}$$

$$
\begin{aligned}
p^0 &: \quad Dy^{(0)} = -dy^{(0)}, \qquad y^{(0)}(0) = c_2,\\
p^1 &: \quad Dy^{(1)} = -dy^{(1)} + ex^{(0)}y^{(0)} - fy^{(0)}z^{(0)}, \qquad y^{(1)}(0) = 0,\\
p^2 &: \quad Dy^{(2)} = -dy^{(2)} + e(x^{(0)}y^{(1)} + x^{(1)}y^{(0)}) - f(y^{(0)}z^{(1)} + y^{(1)}z^{(0)}),\\
&\ \ \vdots
\end{aligned}
$$

$$
\begin{aligned}
p^0 &: \quad Dz^{(0)} = -gz^{(0)}, \qquad z^{(0)}(0) = c_3,\\
p^1 &: \quad Dz^{(1)} = -gz^{(1)} + hx^{(0)}z^{(0)} + iy^{(0)}z^{(0)}, \qquad z^{(1)}(0) = 0,\\
p^2 &: \quad Dz^{(2)} = -gz^{(2)} + h(x^{(0)}z^{(1)} + x^{(1)}z^{(0)}) + i(y^{(0)}z^{(1)} + y^{(1)}z^{(0)}),\\
&\ \ \vdots
\end{aligned}
$$

Consequently, solving the above equations, the first few components of the homotopy perturbation solution for the system (19) are derived as follows

$$
\begin{aligned}
x^{(0)} &= c_1 \exp(at),\\
y^{(0)} &= c_2 \exp(-dt),\\
z^{(0)} &= c_3 \exp(-gt),
\end{aligned}
$$

$$
\begin{aligned}
x^{(1)} &= \frac{bc_1c_2}{d}\exp((a-d)t) + \frac{cc_1c_3}{g}\exp((a-g)t)\\
&- \left(\frac{bc_1c_2}{d} + \frac{cc_1c_3}{g}\right)\exp(at),\\
y^{(1)} &= \frac{ec_1c_2}{a}\exp((a-d)t) + \frac{fc_2c_3}{g}\exp(-(d+g)t)\\
&- \left(\frac{ec_1c_2}{a} + \frac{fc_cc_3}{g}\right)\exp(-dt),\\
z^{(1)} &= \frac{hc_1c_3}{a}\exp((a-g)t) - \frac{ic_2c_3}{d}\exp(-(d+g)t)\\
&- \left(\frac{hc_1c_3}{a} - \frac{ic_2c_3}{d}\right)\exp(-gt).\\
&\ \ \vdots
\end{aligned}
$$

Example 5 *Consider the system [Aziz-Aaloui 2006]*

$$
\begin{aligned}
Dx(t) &= a_0x(t) - b_0x^2(t) - \tfrac{v_0x(t)y(t)}{d_0+x(t)},\\
Dy(t) &= -a_1y(t) + \tfrac{v_1x(t)y(t)}{d_1+x(t)} - \tfrac{v_2y(t)z(t)}{d_2+y(t)},\\
Dz(t) &= a_2z(t) - \tfrac{v_3z^2(t)}{d_3+y(t)},
\end{aligned}
\tag{21}
$$

where a_0, b_0, v_0, d_0, a_1, v_1, d_1, v_2, d_2, a_2, v_3 and d_3 are model parameters assuming only positive values, subject to the initial conditions

$$x(0) = c_1 \quad , \quad y(0) = c_2 \quad , \quad z(0) = c_3. \tag{22}$$

Multiplying the first equation by the factor $d_0 + x(t)$, the second equation by the factor $(d_1 + x(t))(d_2 + y(t))$ and the third equation by the factor $d_3 + y(t)$. According to the homotopy given in Eq. (7), the first few components of the homotopy perturbation solution for the system (21) are derived as follows:

$$\begin{aligned}
x^{(0)} &= c_1,\\
y^{(0)} &= c_2,\\
z^{(0)} &= c_3,\\
\\
x^{(1)} &= \Big(a_0 c_1 - b_0 c_1^2 - \frac{v_0 c_1 c_2}{d_0 + c_1}\Big)t,\\
y^{(1)} &= \Big(-a_1 c_2 + \frac{v_1 c_1 c_2}{d_1 + c_1} - \frac{v_2 c_2 c_3}{d_2 + c_2}\Big)t,\\
z^{(1)} &= \Big(a_2 c_3 - \frac{v_3 c_3^2}{d_3 + c_2}\Big)t.\\
&\vdots
\end{aligned}$$

4 Conclusion

In this work, the HPM has been successfully applied to construct approximate solutions for nonlinear systems of differential equations. The method were used in a direct way to study nonlinear ecosystems.

There are some remarks to make here. First, the HPM doesn't require a small parameter in an equation and the perturbation equation can be easily constructed by a homotopy in topology. Second, the HPM provides the solution in terms of convergent series with easily computable components. Third, the results show that the homotopy perturbation solution in example 1 converges to the exact solution and the approximate solution in example 2 is the same approximate solution obtained using Adomian decomposition method and variational iteration method. Fourth, it is clear and remarkable, from example 3, 4 and 5, that the HPM is effective and simple in order to solve nonlinear systems, specifically predator-prey systems with 2 or 3 populations. It can be easily generalized to any finite populations number.

References

[Abbasbandy 2006a] Abbasbandy, S. (2006) *Homotopy perturbation method for quadratic Riccati differential equation and comparison with Adomian's decomposition method.* Appl. Math. Comp., 172, 485-490.

[Abbasbandy 2006b] Abbasbandy, S. (2006) *Numerical solutions of the integral equations: Homotopy perturbation method and Adomian's decomposition method.* Appl. Math. Comp., 173, 493-500.

[Aziz-Aaloui 2006] Aziz-Alaoui, M.A. (2006) *Complex emergent properties and choas (de-) synchronization.* in M.A. Aziz-Alaoui and C. Bertelle (eds), "Emergent Properties in Natural and Artificial Dynamical Systems", Springer.

[El-Shahed 2005] El-Shahed, M. (2005) *Application of He's homotopy perturbation method to Volterra's integro-differential equation.* Int. J. NonLin. Sci. Mumer. Simulat., 6(2) , 163-168.

[Momani et al. 2008] Momani, S.; and Z. Odibat (2008) *Numerical approch to differential equations of fractional order.* J. Comput. Appl. Math. (in press).

[Odibat et al. 2008] Odibat, Z.; and S. Momani (2008) *Modified homotopy perturbation method: application to quadratic Riccati differential equation of fractional order.* Chaos, Solitons & Fractals (in press).

[He 1999] He, J. (1999) *Homotopy perturbation technique.* Comput. Meth. Appl. Mech. Eng., 178, 257-262.

[He 2000] He, J. (2000) *A coupling method of homotopy technique and perturbation technique for nonlinear problems.* Int. J. Non-Linear Mech., 35(1), 37-43.

[He 2003] He, J. (2003) *Homtopy perturbation method: a new nonlinear analytic technique.* Appl. Math. Comp., 135, 73-79.

[He 2004a] He, J. (2004) *The homtopy perturbation method for nonlinear oscillators with discontinuities.* Appl. Math. Comp., 151, 287-292.

[He 2004b] He, J. (2004) *Comparsion of homtopy perturbation method and homotopy analysis method.* Appl. Math. Comp., 156, 527-539.

[He 2004c] He, J. (2004) *Asymptotology by homtopy perturbation method.* Appl. Math. Comp., 156, 591-596.

[He 2005a] He, J. (2005) *Homotopy perturbation method for bifurcation of nonlinear problems.* Int. J. NonLin. Sci. Mumer. Simulat., 6(2), 207-208.

[He 2005b] He, J. (2005) *Periodic solutions and bifurcations of delay-differential equations.* Physics Letters A, 374(4-6), 228-230.

[He 2005c] He, J. (2005) *Application of homotopy perturbation method to nonlinear wave equations.* Chaos, Solitons & Fractals, 26(3), 695-700.

[He 2005d] He, J. (2005) *Limit cycle and bifuraction of nonlinear problems.* Chaos, Solitons & Fractals, 26(3), 827-833.

[He 2006a] He, J. (2006) *Homotopy perturbation method for solving boundary value problems.* Physics Letters A, 350(1-2), 87-88.

[He 2006b] He, J. (2006) *Some asymptotic methods for strongly nonlinear equations.* Int. J. Modern Physics B, 20(10), 1141-1199.

[Siddiqui et al. 2006a] Siddiqui, A.; Mahmood, R.; and Q. Ghori (2006) *Thin film flow of a third grade fluid on moving a belt by He's homotopy perturbation method.* Int. J. NonLin. Sci. Mumer. Simulat., 7(1), 7-14.

[Siddiqui et al. 2006b] Siddiqui, A.; Ahmed, M.; and Q. Ghori (2006) *Couette and poiseuille flows for non-Newtonian fluids.* Int. J. NonLin. Sci. Mumer. Simulat., 7(1), 15-26.

Part II

Swarm intelligence and neuronal learning

Multi Objective Optimization Using Ant Colonies

Feïza Ghezail[1,2], Henri Pierreval[1], and Sonia Hajri-Gabouj[2]

[1] LIMOS UMR CNRS 6158
IFMA, Institut Français de Mécanique Avancée
Campus des Cézeaux, BP 265
F-63175 Aubière cedex, France

[2] URAII, INSAT, Institut National des Sciences Appliquées et de la Technologie
Centre urbain nord, BP 676
1080 Tunis, Tunisie

Summary. Ant colonies are more and more used to solve various optimization problems, such as scheduling problems. In practice, it is often necessary to take into account several objectives in the optimization procedure. In this respect, ant colonies algorithms have to be adapted to be able to find a set of good solutions that cover in the best way the various regions of the Pareto front. In the following, we suggest an approach that can be used to address optimization problems with a few objectives. We will focus on visibility and desirability issues to favour diversity of solutions in the Pareto front. Further research direction will also be highlighted.

Key words: Multi objective, ant colony.

1 Multi Objective Ant Colony Optimization

Several articles related to multi objective ant colony optimization have already been published. Gravel et al. [Gravel et al. 2002] address a multi objective scheduling problem. They use multiple visibility measures that they combine to determine the global visibily of an ant. The global update of the pheromone is based on the best solution found, at each ant cycle, using a function aggregating the three objectives handled. A sequencing problem is presented by McMullen [Mc Mullen 2001]. Two objectives are considered: to minimize setups and stability of material usage rate. Only one visibility measure is used; the pheromone is updated according to the smallest Euclidean distance computed. Doerner et al. [Doerner et al. 2006] describe a multi objective project portfolio selection problem. The update of the pheromone trail is based on the two bests solutions obtained at each run for each objective handled. A Pareto archive is used to store the non dominated solutions. A reliability optimization problem is addressed by Zhao et al. [Zhao et al. 2007]. The two visibility measures are reduced to a single one using a ratio. Pinto and Barãn [Pinto et al. 2005] solved a multicast routing problem using two different algorithms: a Multi-objective Ant Colony Optimization Algorithm and a Multi objective Min-Max Ant System. A Pareto archive is used to update the pheromone trail. However, methods that would aim at favouring diversity of solutions in the Pareto set are not described. In the approach presented next, emphasis is put on searching for this diversity.

2 Muti objective Ant Colony Approach

2.1 General framework

Step 1
Initialize the pheromone trail and initialize the Pareto set to an empty set

Step 2
For each ant, compute the visibility measures associated with each objective, so as to select the successive nodes according to visibility and pheromone amount, and locally update the pheromone trail until all nodes selected

Step 3
Try to improve the obtained solutions using a local search

Step 4
Evaluate the obtained solutions according to the different objectives and update the Pareto archive with the non dominated ones and reduce the size of the archive if necessary

Step 5
Identify several best solutions according to the different objectives considered

Step 6
Globally update the pheromone, according to the best solutions computed at step 5

Iterate from Step 2 until the maximum of iterations is reached.

Fig. 1. General procedure of the proposed algorithm

Dealing with several objectives in ant colonies that use principles proposed by Dorigo and Gambadella [Dorigo et al. 1997] necessitates to answer three questions: (1) how to globally update pheromone according to the performance of each solution on each objective, (2) how does a given ant locally selects a path, according to the visibility and the desirability, at a given step of the algorithm (3) how to build the Pareto front. Figure 1, summarized the main steps of such an algorithm.

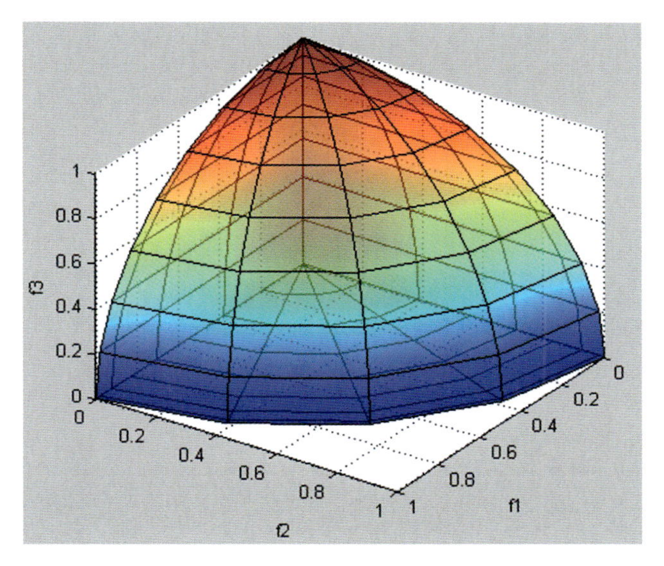

Fig. 2. Search directions for a maximization problem

2.2 Pheromone and desirability

Each time ants select successive paths to construct a solution, the pheromone trail τ_{ij} of each segments (i, j), cant be locally updated according to:

$$\tau_{ij}(t+1) = (1-\rho)\tau_{ij}(t) + \rho\tau_0 \tag{1}$$

ρ $(0 < \rho < 1)$ is a persistence factor and τ_0 is a constant. This local update of the pheromone is used to evaporate some quantitiy of pheromone to avoid a premature convergence of the algorithm.

Then, at the end of an iteration, every ant has found a solution s and the pheromone trails have to be globally updated on the basis of the performance achieved on the u objectives $f1, ..., fu$. To select the set of paths for which pheromone has to be reinforced, we determine, from the set of available solutions, those that have yield the best results on w linear combinations of the objectives:

$$F^p(s) = \lambda_1^p f_1(s) + \lambda_2^p f_2(s) + ... + \lambda_u^p f_u(s), \quad p = 1, ...w. \tag{2}$$

These functions characterize w search directions, which can be determined through the w vectors: $(\lambda_1^1, ..., \lambda_u^1), (\lambda_1^2, ..., \lambda_u^2), ..., (\lambda_1^w, ..., \lambda_u^w)$ as illustrated in Figure 2 in the case of three objectives. This incites the algorithm to explore systematically distinct areas, so as to favor the diversity of solutions in the Pareto set [Siarry et al.]. Let s^p_{best} be the solution that yields the best results with F^p. Then the pheromone of each of the $n-1$ segments (i,j) of the corresponding path is reinforced in a minimization problem as follows,

$$\tau_{ij}(t+n) = (1-\rho)\tau_{ij}(t) + \rho\tau^p_{ij}, \quad p = 1,...,w. \tag{3}$$

$$\Delta\tau^p_{ij} = \begin{cases} \dfrac{1}{F^p(s^p_{best})}, & \text{if } (i,j) \in \text{ best solution according to the uple p} \\ 0 & \text{otherwise} \end{cases} \tag{4}$$

Where ρ $(0 < \rho < 1)$ is a persistence factor, t the current discrete time and n the number of nodes of a path. Let us note that this approach is adapted if the number u of objective is low.

2.3 Visibility

In addition to the pheromone quantity, ants are guided by a proximity measure called *visibility*. Since several objectives are considered [Liao et al. 2007], several visibility measures can be used, depending on the problem. Visibility values can be stored in a matrix connecting each node i to each node j. For example, in [Gravel et al. 2002], a visibility measure η^c_{ij} is defined for each objective c, $(c = 1,...,u)$ and combined. Then, each ant k, $(k = 1,...,m)$ that leaves node i selects the next node j to be visited according to the probability given in (4), where q is a randomly generated variable and q_0 is a parameter, such that $q \geq 0$, $q_0 \leq 1$. α and β_c are the control parameters and $tabu_k$ is a memory list used to avoid reselection of nodes already chosen by each ant k.

$$p^k_{ij} = \begin{cases} \dfrac{[\tau_{ij}(t)]^\alpha \prod\limits_{c=1,...,u} \left[\dfrac{1}{\eta^c_{ij}}\right]^{\beta_c}}{\sum\limits_{l \notin tabu_k} [\tau_{il}(t)]^\alpha \prod\limits_{c=1,...,u} \left[\dfrac{1}{\eta^c_{il}}\right]^{\beta_c}} & if j \notin tabu_k \\ 0 & \text{otherwise} \end{cases} \tag{5}$$

2.4 Multi objective local improvment

According to Hu et al. [Hu et al. 2005], an important weakness of the Ant Colony algorithm is that the search may fall into a local optimum. An improvement function, multi objective in our case is useful to enhance the ACO performance. A possible approach consists in selecting p solutions from those generated by the algorithm and in modifying them with some elementary mofifications (e.g. 2-OPT for a scheduling problem). Then, the best l ones are stored in the Pareto archive. To select the l best solutions, l directions are defined to favor the less populated area of the current Pareto front, so as to improve the diversity of the proposed solutions to the decision maker.

2.5 Pareto selection

The set of non-dominated solutions is stored in an archive. During the optimization search, this set, which represents the Pareto front, is updated [Loukil et al. 2005]. At each iteration, the current solutions obtained are compared to those stored in the Pareto archive; the dominated ones are removed and the non dominated ones are added to the set. The size of this set needs to be kept reasonable, which may imply to sometimes remove non dominated solutions. As suggested for multi objective genetic algorithms, to preserve the diversity of the set, solutions belonging to the most populated areas can be removed first.

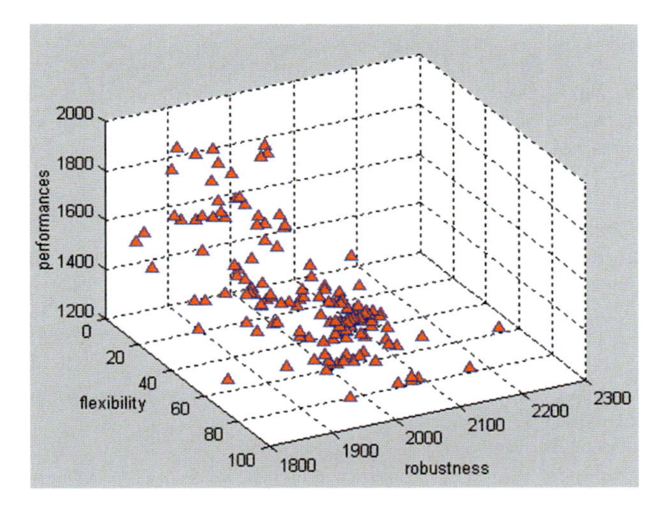

Fig. 3. The Pareto set distribution

3 Example

We tested the proposed approach for a single machine multi objective problem related to a printing shop. Each product has a size and a printing label that needs different ink colours, which induce constraints about the tool to be used (mandrel) and on the sequence of jobs. We consider groups of jobs having the same size to be scheduled. We are interested in minimizing a performance function (based on ink changes and total tardiness), a robustness measure (based on a regret in case of machine breakdown), and a flexibility measure that quantifies possible lost of performance if the tool was not available (see in [Ghezail et al. 2005] for more details). At each iteration of the algorithm, the number of solutions in the Pareto set staid low, so there was no need to eliminate solutions in the archive. The resulting Pareto set allows the decision maker to select the most suited schedule, according to the operating conditions of the workshop.

4 Conclusion

The method proposed here aims at improving the diversity in the Pareto set, so as to offer different types of compromises to the decision makers. The key principles are simple to implement in an ant algorithm, but they are mainly suited for problems with few objectives. One remaining important research issue is related to the adaptation of visibility principles to multi objective problems. Currently, the different visibility measures are aggregated in a single one using weights or a ratio. An interesting research direction would consist in adapting the influence of each visibility measures depending on what solutions are available in the Pareto set.

References

[Doerner et al. 2006] Doerner, K.F.; Gutjahr, W.J.; Hartl, R.F.; Strauss, C.; and C. Stummer (2006) *Pareto ant colony optimization with ILP preprocessing in multiobjective project portfolio selection.* European Journal of Operational Research, 171, 830-841.

[Dorigo et al. 1997] Dorigo, M.; and L.M. Gambadella (1997) *Ant colonies for the travelling salesman problem.* BioSystems, 43, 73-81.

[Dréo 2003] Dréo, J. (2003) *Adaptation de la méthode des colonies de fourmis pour l'optimisation en variables continues.* Application en génie biomédical. PhD Dissertation, University of Paris 12, France.

[Ghezail et al. 2005] Ghezail, F.; Hajri-Gabouj, S.; and H. Pierreval (2005) *A multiobjective Ant Colony Approach for a Robust Single Machine Scheduling problem.* Proc. of International Conference on Industrial Engineering and System Management, Marrakech, Maroc, 479-488.

[Gravel et al. 2002] Gravel, M.; Gagné, C.; and W.L. Price (2002) *Algorithme d'optimisation par colonies de fourmis avec matrices de visibilité multiples pour la résolution d?un problème d'ordonnancement industriel.* INFOR, 40(3), 259-276.

[Hu et al. 2005] Hu, Y.-H.; Yan, J.-Q.; YE, F-F.; and J.-H. Yu (2005) *Flow shop rescheduling problem under rush orders.* Journal of Zhejiang University SCIENCE 6A(10), 1040-1046.

[Liao et al. 2007] Liao, C-J.; and H.-C. Juan (2007) *Ant Colony optimization for single machine tardiness scheduling with sequence-dependent setups.* Computers and Operation Research, 34(7), 1899-1909.

[Loukil et al. 2005] Loukil, T.; Teghem, J.; and D. Tuyttens (2005) *Solving Multiobjective Production Scheduling using metaheuristics.* European Journal of Operational Research, 161, 42-61.

[Mc Mullen 2001] Mc Mullen, P.R. (2001) *An ant colony optimization approach to addressing a JIT sequencing problem with multiple objectives.* Artificial Intelligence in Engineering, 15, 309-317.

[Pinto et al. 2005] Pinto, D., Barãn, B. (2005) *Solving Multiobjective Multicast Routing Problem with a new Ant Colony Optimization approach.* In proceeding of IFIP/ACM Latin-American Networking Conference LANC'O5, Cali, Colombia.

[Siarry et al.] Siarry, P.; and Y. Collette. (2002) *Optimisation multiobjectif.* Eyrolles, Paris, France.

[Zhao et al. 2007] Zhao, J.H.; Liu, Z.; and M.T. Dao. (in press). *Reliability optimization using multiobjective ant colony system approaches.* Reliability Engineering and System Safety, 92(1), 109-120.

Self-Organization in an Artificial Immune Network System

Julien Franzolini and Damien Olivier

LITIS - University of Le Havre
25 rue Ph. Lebon, BP 540
76058 Le Havre cedex - France
julien.franzolini@univ-lehavre.fr, damien.olivier@univ-lehavre.fr

Summary. Artificial Immune System field uses of the natural immune system as a metaphor for computational problems. The immune system exhibits a highly distributed, adaptive and self-organizing behavior. Furthermore it can learn to recognize shapes with its adaptive memory. The approach explored is inspired by an immune network model like Stewart-Varela's model. This model is designed to better understand the memory and cognition properties. Emergent antibodies configurations are studied on this immune network model; indeed these configurations seem like cellular automata configurations and appear with self-organizing properties.

Key words: AIS, immune network, idiotypic network, self-organisation

1 Introduction

The immune system is a natural complex system which protects all vertebral organisms. It is able to recognize foreign molecules and to learn to detect this foreign agent more quickly and more effectively. This system has many proprieties: it is adaptive, self-organizing, distributed and robust. All these capabilities interest the computer sciences for various domains : clustering, intrusion detection and optimization. This paper explains fundamental biological theories used in artificial immune systems and in a second time it presents a model of immune system based on immune network theory.

2 The Immune System

2.1 General Description of the Immune System

The immune system is really a complex system due to the number of actors and the multiple interactions involved in an immunizing response. This system protects the body against pathogen agents by various collective mechanisms. These mechanisms termed immunity give the state of protection against a foreign agent called antigen.

The immune system is composed of lymphocytes which are known under the name white blood cells, more exactly B and T cells. These B cells help the process of antigens recognition by secreting antibodies corresponding to an antigen. Antibodies can fix antigens by a complementary shape Fig.1.

Antigens are majors actors because they are the attractor of the immunizing response, without these antigens the system does not engage a response. There is two type of immune response:

- The primary response comes from the natural or innate immunity, this response is provoked when an antigen is encountered for the first time. The B lymphocyte specific to the antigen generates many antibodies oriented to the antigen shape. With this first action the antigens can be more easily destroyed.
- The secondary response called acquired or specific is occurred after a similar re-infection and not compulsory with an identical shape. The system can yet produce more quickly and more massively specific antibodies by memorizing the original shape of the past encounter [Timmis et al. 2000]. This memorization of antigens is the occasion of an immunological debate to explain this adaptive memory .

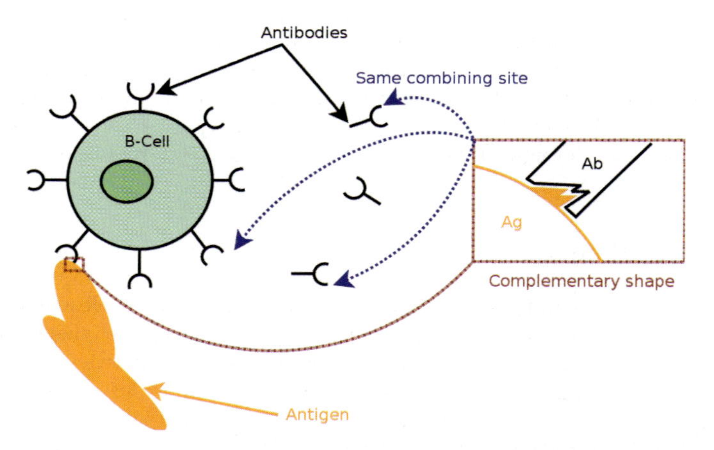

Fig. 1. Affinity Between a B Lymphocyte, its Antibodies and an Antigen.

Many models try to explain how the immune system reacts but theories are in contradiction, as the self-nonself theory and danger theory and some other theories are complementary as the clonal selection theory with the others then it is necessary to give a brief review of these different theories.

2.2 Clonal Selection Theory

When an immune response is engaged, lymphocytes are stimulated to proliferate and secrete their free antibodies corresponding to the antigen. During this proliferation which is realized with cloning, some mutations are made on the new B-cells. These mutations called hyper mutations somatic improve the capacity to fix the pointed antigen. A B-cell and its antigen are specific and code for the same shape. More the affinity of an antibody and an antigen is important more the B-cell is cloned. This phenomenon is known as the maturation of immune response [de Castro et al. 2000] and it is one of immune learning mechanism. The cell with high stimulation (high antigens affinity) creates with cloning to high living cells called memory cells, which are kept in the body during a long time to generate a secondary immune response. Such a minority of B lymphocytes can recognize one specific antigen and is activated by clonal selection then this proliferation increases the specific answer. This principle is nearly of natural selection used in evolutionary algorithm but it does not explain all memory mechanisms and still less how the system recognizes the foreign agents.

2.3 Self and Nonself Discrimination

The immune system can give an immune response and improves it by the clonal expansion but how the system doesn't recognize self antigen (or self agent)? This question is the beginning of an enlivened discussion between immunologists, and involves several interaction of lymphocytes. But the most used immunological explication is the *thymic negative selection of T-cells* [de Castro et al. 2003]. These T lymphocytes which finish their maturation in thymus became mature and are introduced in blood if they don't have affinity with self antigen. With this action, the domain recognition of T-cells maps the non-self universe. The system does not recognize non-self molecules as water, food, stomach bacteria... But these molecules or these organisms are foreign to the body. This problem gives an alternate hypothesis: Danger Theory.

2.4 Danger Theory

The system has the ability to respond to foreign agent but only pathogenic and not only non-self. The danger theory explain that the system recognizes only dangerous invaders. This response is induced by a danger signal (ejection of molecules), when a cell is in stress or killed. In this theory, there are many danger signals not only in secreting molecules, and this different type drives the immune response [Aickelin et al. 2003]. This theory is a new way to use immune system for biological metaphor in particularly for intrusion detection system. All these variant theories are insufficient to explain the antibodies coverage of the diversity of dangerous shapes.

2.5 Immune Network Theory: Idiotypic Network

Over all this theory and hypothesis, the most important for this work is the immune network theory, which tries to explain the immune organization of the antibodies distribution. This theory is based on a theory established by Niels Jerne (Nobel Price for his work) that the lymphocytes are able to enter in interactions . This theory suggests that this interactions network can grow up unless antigens stimulation, to make the antibodies distribution. This theory reposes on a first hypothesis gived by Coutinho that the immune repertory is complete to be able to recognize all various antigens. This hypothesis was completed by Jerne: if the immune repertory is complete, antibodies of the same body have to enter in interaction by their complementary combining site, and so an antibody can provoke an immune response against an other antibody.

The two hypotheses give the theory of a development in network called the idiotypic network. It shows how the system can have sufficient diversity to recognize unknown antigen. This stimulation network Fig.2 maps the universe of possible shapes.

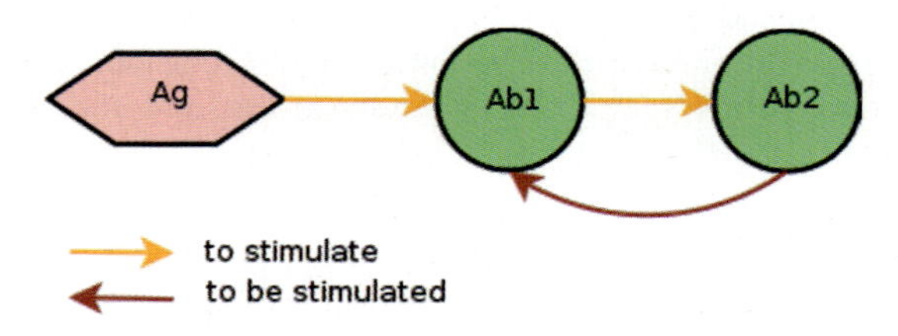

Fig. 2. Ab/Ag and Ab/Ab Stimulations [de Castro et al. 2003]

3 An Immune Network Model

Generally, the interaction from idiotypic network is simulated with the Farmer et Al. [Timmis et al. 2001] equation which describes the antibody interaction between other antibodies, antigens and formulate its death rate :

$$
\begin{aligned}
Stimul = c[(antibodiesrecognize) - (I'mrecognized) \\
+(antigenrecognize)] - (death)
\end{aligned}
$$

A development in networks is chosen, cause of its different properties as self-organization. This work relies on a research from Stewart and Varela [Stewart 1994] where they show cognitive properties from an idiotypic network model. They have tried to make a mathematic model but it was highly non-linear and so not workable. They create so a computer model, where the idiotypic network is symbolised on a form space. In a form space, one dimension corresponds to one stereochemic characteristic, which describes the combining site. De facto, the distance of two entities in this space represents their characteristics differences Fig.3.

Two complementary molecules (Ag/Ab or Ab/Ab) able to fix each other are close in the space. In their model, antibodies were recruited if their stimulation (depend on affinity whit other complimentary antibodies) belongs to a recruitment window.

3.1 General Principle of the Conceived Model

The Stewart-Varela affinity calculation between two antibodies is kept in this new model (equation 1), but the calculation of the stimulation received by an antibody is

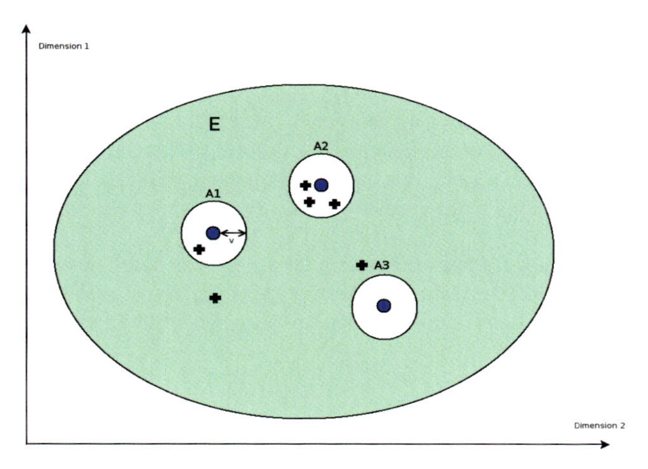

Fig. 3. Two Dimensions Form Space.

different (equation 2). Not only two types of complementary antibodies regrouped by plan are considered but n types so n plans. The stimulation undergoes by an antibody is the sum of all antibodies affinities of all other complemantary plans.

$$m_{ij} = e^{-d_{ij}^2} \tag{1}$$

m_{ij} is the affinity and d_{ij} the distance between two antibodies i and j in the form space. An Euclidian distance is used but for example an Manhattan distance could be considered.
For a given plan p:

$$h_i = \sum_{k=1}^{k<nbPlan,k\neq p} \sum_{j=1}^{j<N_k} m_{ij} \tag{2}$$

h_i is the stimulation received by an antibody i belonging to the plan p, $nbPlan$ is the number of plans and N_k the antibodies number for a plan k.

The calculation of the distance is realized in a form space which has the topology of a tore. The simulation begins by an antibody introduction from a random plan in the center of the space. After, antibodies are randomly subjected in the recruitment to be introduced into the system. If one corresponds (its stimulation belongs to the window of recruitment $h_i \in]b_{min}; b_{max}[$) the search for a candidate stops and the found antibody is introduced into the system .

The goal of this model is to develop itself in self-organized criticality, to maintain the criticality after each introduction, all antibodies stimulations are recalculate to be sure that all antibodies belong to the recruitment window. An antibody is rejected from the system if its stimulation is not sufficient or too high.

3.2 Details Model

The details of the principal phase called development are given by the algorithm 1. This algorithm ends if the system becomes stable. The stability is defined by the fact that no antibody can be recruited in the system : the stimulation of all the possibilities of shapes doesn't agree with the recruitment window and in the previous phase all the stimulations were verified, so the system does not evolve any more.

In these algorithms, antibodies and lymphocytes are the same entities because they code for the same characteristics (attributes) and in this model there is no clonal expansion. The lymphocytes concentration for a point of the form space is 0 or 1.

begin
  Introduction of a central lymphocyte ;
  Evolution ← true ;
  while *Evolution* **do**
    Evolution ← recruitmentLymphocyte() ;
    Calculate lymphocyte interactions in the different plans ;
    Deletion of a unique lymphocyte which don't agree with the recruitment window ;
    while *deletion of one lymphocyte* **do**
      Calculate lymphocyte interactions in the different plans ;
      Deletion of a unique lymphocyte which don't accord with the recruitment window ;
    end
  end
end

Algorithm 1: Principle of Development

Problems of collapse and death of systems can be encored. This fact comes from the suppression of all antibodies which cannot be maintained, it is the reason for the deletion of a unique antibody, particularly the one which undergoes most pressure (having the highest stimulation).

To give freedom to the system, all plans and positions in the form space are randomly selected. The critical state of the system is maintained by continuing the elimination of the lymphocytes as long as their stimulations do not agree. This phenomenon is called distribution of avalanches.

4 Results and Model Properties

4.1 A Self Organizing System

Before running the system, it is necessary to choose adequate bounds. With arbritary bounds, no developments emerge and on the contrary the lymphocytes fill the

```
while candidate exists do
    Select at random of a Lymphocyte among the population of candidate ;
    Select at random of a plan ;
    while plan opposed exists do
        Calculation of the lymphocyte interactions with each opposed plans;
        if can be recruit then
            insertion of antibody in the system;
            return true ;
        end
        Select at random of a new opposed plan ;
    end
    Select at random of a new candidate ;
end
return false ;
```

Algorithm 2: Recruitement Mechanism

form space randomly without organization.

Fig.4 shows a development with self-organization in two dimensionnal form space (better for visualization). This form space has a size of 40 possibilities by dimension and the topology of the space is a tore. Each color represents a complementary plan. The system is started with the two bounds: $b_{min} = 2.10^{-10}$ and $b_{max} = 22.10^{-8}$; and with four complementary plans. At the beginning the system evolves from the first central lymphocyte and then enters in a organization phase to become stable at the iteration 898.

Fig.5 gives the evolution of lymphocytes by plan and the global population for the previous example. For an other run, the obtained system has the same morphology but it has not an identical lymphocytes configuration due to the random submission to the recruitement. An iteration corresponds to a whole development phase, this phase contains a single introduction but no or many deletions. For each iteration the system has a critical configuration (All stimulations belong to the window).

4.2 A Behavior Dependent on Bounds

The system can develop or not according to the bounds, and have a self-organized behavior or not. That seems like cellular automata where the behavior depends on the complexity class [Wolfram 1993]. One of major results is that the bounds condition the development and the characteristics of the system: lymphocytes activities (recruitement or death), convergence time to a stable state or no stabilization and morphologies of emergent lymphocytes organization.

The example Fig.6 exhibits tree different systems with the same recruitement window: $b_{min} = 36.10^{-9}$, $b_{max} = 36.10^{-8}$. But the first has 2 plans and a size of 50, the second has 2 plans and a size of 100 and the last has 5 plans for a size of 50. The size and the number of plans have no influences on the system behavior.

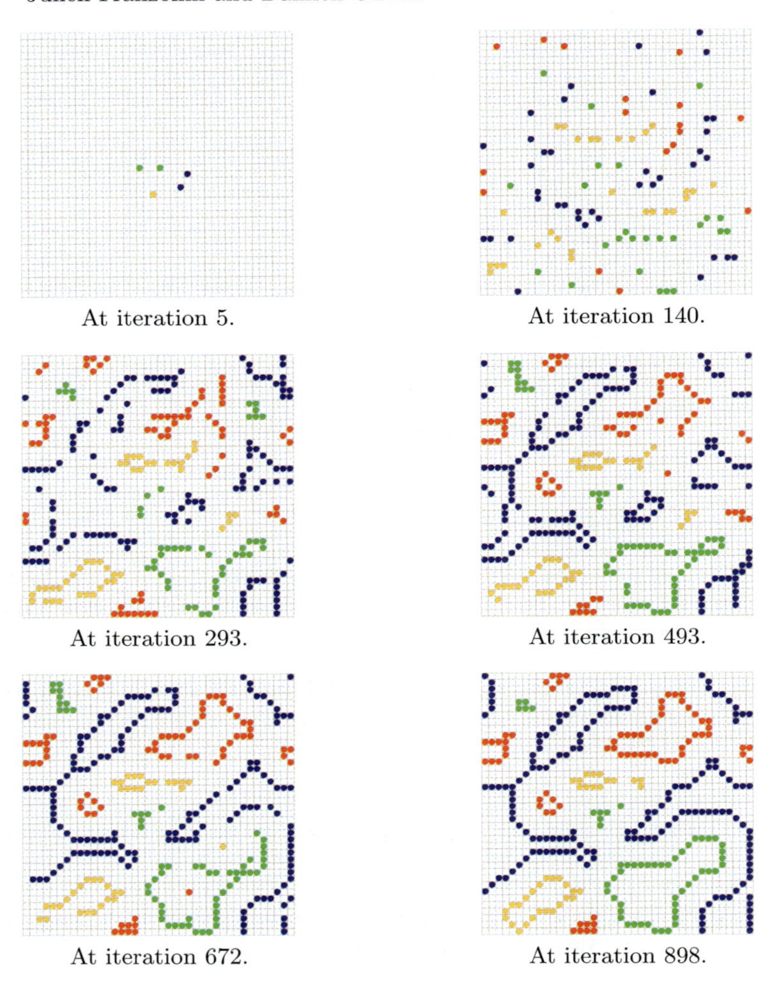

Fig. 4. An Immune Network Development

4.3 Self-organized Criticality

One of caracteristics of the self-organized criticality (SOC) is the macroscopic behaviour. The self-organized critical systems displays a spatial scale-invariance (or time-invariance) characteristic of the critical point of a phase transition, but, unlike the latter, in SOC these features result without needing to tune control parameters to precise values [Bak 1996].

The system exhibits auto-organized behaviors as the lymphocytes avalanches (lymphocytes deletions). To study this phenomenon, avalanches are ploted in double logarithm scales functions of the avlanches size and the number of occurences by size. The result Fig.8 for $b_{min} = 2.10^{-8}$ and $b_{max} = 2.10^{-7}$ is not a power law which

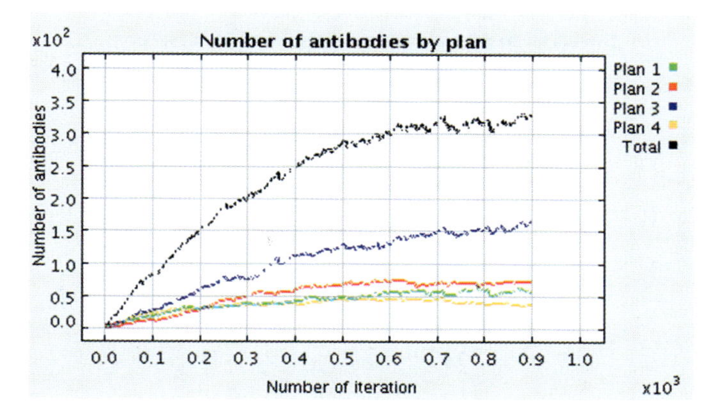

Fig. 5. Lymphocyte Evolution by Plan during the Development

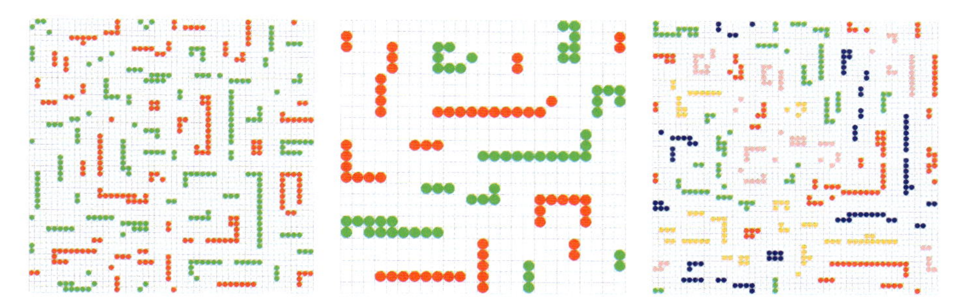

Fig. 6. Identical Morphologies with the Same Bounds.

is the most frequent scaling laws that describe the scale invariance found in many natural phenomena, but anyway a distribution.

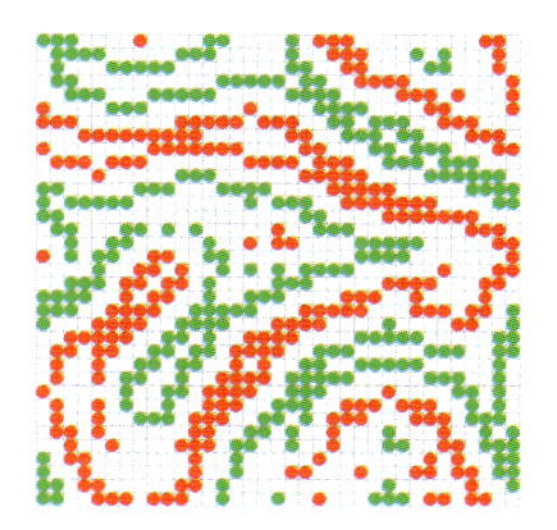

Fig. 7. Unstable System with Lymphocytes Avalanches.

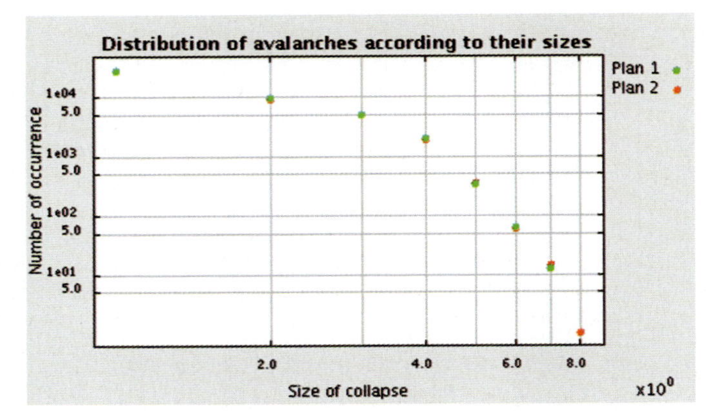

Fig. 8. Avalanches Distribution Functions of Size.

4.4 Antigens Reactions

To test the antigens reactions, a form is introduced in the system by forcing. This antigen does not undergo the phase of maintain or recruitement. An antigen is introduced Fig.9 (square shape on the figure). This introduction changes the stimulations of the other lymphocytes and modifies the lymphocytes configuration.

Another aspect is that more the exposure of the antigen is long more the system is modified and conversely a little exposure does not distort the configuration. The exposure duration is corresponding to the number of iterations. With the theory [Stewart 1994] according to the network morphology conditions the memory, it is supposed that the time of exposure amplifies or not the learning of the immune network by deformation of the spatial configuration.

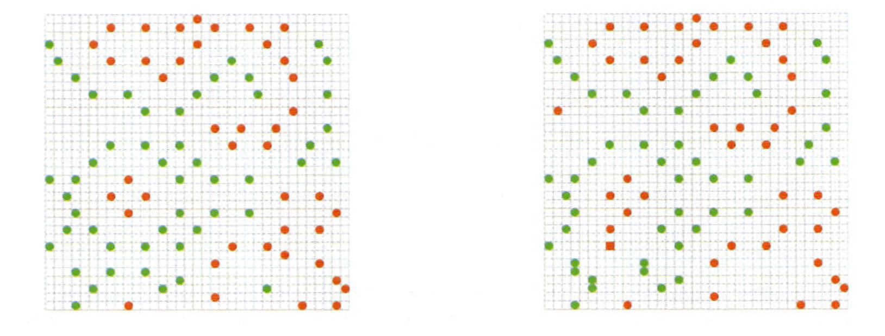

Fig. 9. Antigen Introduction.

4.5 Comportements Maps

One of this model problems is how to have bounds which would give self-organized properties, to find more easily valid bounds (conceivable development). Comportements maps are drawed by making vary bounds. These maps represent the number of iterations to converge for a death, a stable state or a no stabilization.

The blue color Fig.10 indicates the system death (system whitout lymphocytes), the green one represents a system still alive or stable and the black shows the impossible developments. A map is built for a given size and a given number of plans. The luminous intensity of the blue or the green indicates the speed of convergence to arrive at a given behavior, indeed at every iteration a test is realized to know if the system is empty or not, if it is stable or in development. An average on several launches with identical bounds is realized to give an usable map.

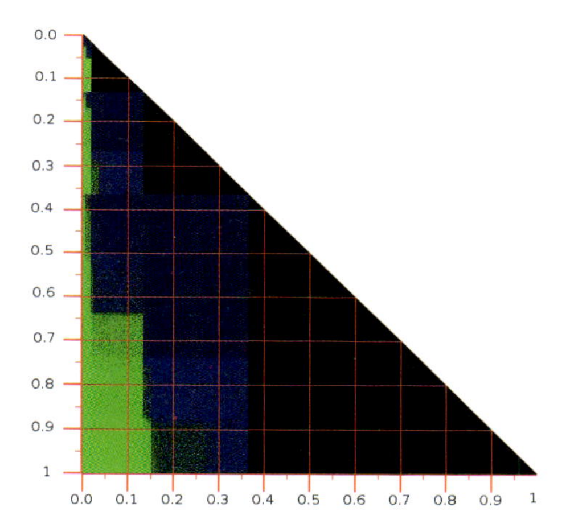

Fig. 10. An Comportements Map with $b_{min}, b_{max} \in [0, 1]$, 2 Plans and a Size of 25 by Plan.

Conclusion

This model allows to approach better the different interactions which form the idiotypic network and so to understand how these very simple and multiple interactions can generate a memory and realize a learning. One of the perspectives is to manage the behavior by bounds (maybe with comportements maps) to realize an artificial memory able to develop in self-organization and with a critical state to be adaptative.

References

[Aickelin et al. 2003] Aickelin, U.; Bentley, P.; Cayzer, S.; and J. Kim (2003) *Danger Theory: The Link between AIS and IDS?.* Proceedings of the 2nd International Conference on Artificial Immune System (ICARIS 2003), 147-155.

[Bak 1996] Bak, P. (1996) *How Nature works : the science of self-oganized criticality.* Springer-Verlag.

[de Castro et al. 2000] de Castro, L.N.; and F. J. Von Zuben (2000) *An Evolutionary Immune Network for Data Clustering.* in Proceedings of the IEEE SBRN'00 (Brazilian Symposium on Artificial Neural Networks), 84-89.

[de Castro et al. 2003] de Castro, L.N.; and J. Timmis (2003) *Artificial immune systems as a novel soft computing paradigm.* Soft Computing(7), 526-544.

[Stewart 1994] Stewart, J. (1994) *Un système cognitif sans neurone : les capacités d'adaptation, d'apprentisage et de mémoire du système immunitaire.* Intellectica 1994/1, 18, 15-43.

[Timmis et al. 2000] Timmis, J.; Neal, M.; and J. Hunt (2000) *An Artificial Immune System for Data Analysis*, Biosystems(55), 143-150.

[Timmis et al. 2001] Timmis, J.; and T. Knight (2001). *Artificial Immune Systems: Using the immune system as inspiration for data mining.* in Hussein A. Abbas, Ruhul A. Sarker and Charles S. Newton (eds), *Data Mining: A Heuristic Approach*, Chapter XI, 209-230, Idea Group Publishing.

[Wolfram 1993] Wolfram S. (1993) *Statistical Mechanics of Cellular Automata.* Review of Modern Physics (55), 601-644.

On Adapting Neural Network to Cellular Manufacturing

Dania A. El-Kebbe and Christoph Danne

Paderborn University / Heinz Nixdorf Institute
Fürstenallee 11
33102 Paderborn, Germany
elkebbe@uni-paderborn.de

Summary. This work gives an overview of the neural network based approaches that have been developed to solve the part-machine grouping problem related with the introduction of a cellular manufacturing layout. It also proposes and discusses extensions that should be made to this tool in order to overcome its drawbacks.

Key words: Neural networks, Group Technology, Cellular Manufacturing, Part Machine Grouping.

1 Introduction

In today's business environment, manufacturing companies are constantly searching for improved methods to be able to produce high quality products at low costs. Among the organizational approaches that emerged in past decades, Group Technology (GT) has received considerable attention. In essence, GT seeks to identify similarities among parts and the exploit these similarities in different stages of the production process. The idea is to find groups of parts that to some extend share the same design and manufacturing features and then take advantage of this knowledge in many stages of the manufacturing cycle, like engineering design, process planning, production planning and the production process itself.

The term Cellular Manufacturing (CM) describes a major application of the GT philosophy in the production stage and seeks to bring some of the benefits of mass production to less repetitive job shop manufacturing systems. In contrast to the traditional job shop layout, where similar machines and labor with similar skills are located together, the Cellular Manufacturing approach seeks to form independent cells of machines and labor that are capable of producing one or more groups of similar parts (part families) independently, by that decomposing the manufacturing system into subsystem. The adoption of Cellular Manufacturing has proven to yield some considerable benefits, like reduced setup times, reduced lead times and less material handling.

As the exploitation of similarities among parts forms the basis for GT applications, the detection of these similarities, referred to as *part family formation* and *part classification*, is a prerequisite for any such application. The implementation of a cellular manufacturing system furthermore poses the *part-machine grouping* problem, which describes the task of forming ideally independent machine cells that are able to produce one or more part families autonomously. In practice, it is not always possible to form completely independent machine cells, because the requirement of the same machine by at least two parts from different part families results in inter-cell movements. Such machines are called *bottleneck machines*, the corresponding parts are referred to as *overlapping parts.* A wide range of different approaches has been proposed in past decades to address these problems, including methods based on graph theory [Rajagopalan et al. 1975], matrix sorting [King 1980, King et al. 1982], mathematical integer programming [Kusiak 1987] and expert systems [Kusiak 1988]. Most of these approaches have several drawbacks in common, like computational inefficiency when working on industry size datasets and inflexibility when new parts are introduced.

Neural networks offer some unique capabilities that make them extremely suitable for this type of applications. The fact that they can learn by experience to recognize patterns combined with their ability to generalize the knowledge they have obtained are the outstanding advantages that justify the use of neural networks for Cellular Manufacturing applications. Furthermore, when used for clustering, they can classify newly introduced inputs without the need to cluster the entire dataset again. Another advantage is that neural networks are able to handle incomplete data accurately, thus making them useful in real-world applications that have to cope with such information.

2 Applications of Neural Networks to Cellular Manufacturing

Neural networks have shown to be a promising tool to address the problems related with autonomic computing, especially such complex tasks with a great number of input elements, because of their greater flexibility and ability to learn from experience and generalize their acquired knowledge to recognize new input patterns. In recent years, much research was dedicated to the application of neural networks for group technology applications, as they are suitable for complex clustering tasks due to their pattern recognition and generalization abilities.

In order to use neural networks to group parts into part families and/or machines into manufacturing cells, it is necessary to find a metric to measure similarity and define how the relevant part features can be coded to serve as input for neural networks. While other Group Technology applications like design retrieval systems [Venugopal et al. 1992] analyze the part's design features, almost all approaches for part-machine grouping focus on a part's processing requirements, following an approach called *production flow analysis* (PFA) [Burbridge 1991]. This approach abstains from considering every part feature and focuses on the part's routings, as they imply the part's machine requirements and therefore the relevant information to form independent manufacturing cells. A feasible way to represent machine requirements is via *part incidence matrices*, which are binary-valued matrices consisting of a row for each part and a column for each machine involved. Each entry x_{ij} takes a value of 1 to indicate that part i requires an operation to be performed on machine j, and a value of 0 otherwise. Therefore, each row of the matrix represents the machine requirements of the corresponding part as a binary vector of dimension m, with m being the number of machines considered. These vectors serve as the input for the neural networks.

Part-machine grouping includes the subproblems of forming part families as well as machine cells. Therefore, sequential and simultaneous approaches can be distinguished, dependent on whether they try to solve both problems separately or simultaneously. The neural network based models are sequential approaches, although they can be used to solve both clustering tasks one after another. They can form either part families based on machine requirements (taking the rows of part incidence matrices as their input) or machine cells based on part processing capabilities (taking the columns of part incidence matrices as their input). In the following sections, we assume that part feature vectors form the input and that the network is used to create part families.

2.1 Basic network architectures

From the wide range of existing neural network models, some have proven to be especially practical for solving CM related problems. It may be noted that the overwhelming majority of neural networks that have been developed for this task use unsupervised learning methods. This is due to the fact that supervised learning always requires some knowledge about the clusters that will be formed and a set of training data containing input elements with a known correct output value. This type of information is rarely available for part-machine grouping problems, because typically no information about the correct group formation is known a priori.

Nevertheless, Kao and Moon [Kao et al. 1991] propose a method for part grouping based in a three-layer feedforward network using backpropagation learning (i.e. supervised learning). They try to overcome the lack of training data by arbitrarily selecting some parts (*seed parts*) to be the representatives of the part families and train the network to classify these parts correctly. Subsequently, all remaining parts should be classified by the network to belong to one of these part families. This approach suffers from several shortcomings. First, the task of selecting the seed parts is left to human interaction, thereby depending on a subjective judgment to select parts that are as "distinctive" as possible. This is clearly a weak point, as the selection of representatives for the part families heavily affects the number of part families formed and their characteristics. Second, the training process may be repeated several times with an increasing number of elements in the training set. A newly introduced part that does not fit into any of the existing part families causes the network to create a new part family and start the whole training process again.

Against this background, research has focused on network models using unsupervised learning methods as they make weaker assumptions about the underlying clusters. The basic architecture of all network types considered in the subsequent sections consists of two fully interconnected layers. The input layer contains as many neurons as there are columns in the part incidence matrix (i.e. machines considered) and reads one row of the matrix at a time and passes it to the output layer via weighted connections. Each output layer neuron j represents one cluster, therefore the output is a binary vector with only one entry taking a value of 1 and thus indicating the corresponding part family and values of 0 for all other entries. The output neurons "compete" to respond to the input pattern, computing the weighted sum of their input signals. The winning neuron (i.e. the one with the highest input value) indicates the part family the presented part should be assigned to and updates its weight vector to move closer to the centroid of all elements in this cluster. This is done by transferring a part of the weight of the connections receiving values of 0 at the input layer to the connections that receive values of 1.

This basic mode of operation, called *competitive learning* [Venugopal et al. 1994, Chu 1993], still suffers from various shortcomings when applied in the context of part-machine grouping. The number of part families created has to be specified in advance and there are no means to control the degree of similarity among the part families. In the following sections, we will outline the more recent approaches that try to overcome these limitations by extending the model in several ways.

2.2 Kohonen Self Organizing Feature Maps

Self Organizing Feature Maps (SOFM) [Kiang et al. 1995] may be seen as an extension to the competitive learning model. The special characteristic is a two dimensional output layer, also referred to as the output map or Kohonen layer. The SOFM takes a vector of any dimension as its input and can be used to transform it into a two-dimensional map that can be represented graphically (note that in this case the output neurons do not directly represent part families). Kiang, Kulkarni and Tamb [Kiang et al. 1995] claim that the SOFM performs a dimension reduction of the input patterns to the two-dimensional space while maintaining the topological relations between the elements. Thus, clusters in the higher dimensional input will also appear as clusters in the Kohonen layer. This makes this network particularly useful to be integrated into an interactive decision support system, where the graphical representation of the map can be used by a decision maker to fragment it into the desired number of clusters. In contrast to other neural network based approaches, the SOFM does not perform any grouping of parts itself, it just helps to identify clusters by offering a method to visualize the parts to identify similarities.

The operation of the SOFM is similar to the competitive learning model. The main difference is that not only the weight vector of the winning neuron is updated, but also those of the neurons in a predefined neighbourhood. When an input pattern is presented, for each neuron j in the Kohonen layer, the Euclidean distance of its weight vector to the input pattern:

$$D_j = \sqrt{(x_1 - w_{j1})^2 + (x_2 - w_{j2})^2 + ... + (x_m - w_{jm})^2}$$

is computed. Then, the neuron j^* with the smallest distance is selected and its weight vector is updated to move closer to the input pattern according to the equation

$$w_{j^*}^{new} = w_{j^*} + \alpha(X - w_{j^*})$$

where α is the learning rate of the network and controls the speed of weight adoption. To guarantee convergence, it is proposed to start with a high learning rate and a wide neighborhood and decrease both progressively in time.

The Kohonen SOFM have proven to be efficient for part grouping and are a promising approach to be integrated into interactive decision support systems where the final clustering is done by a domain expert.

2.3 Adaptive Resonance Theory

Similar to the competitive learning model, the Adaptive Resonance Theory (ART) [Dagli et al. 1995] seeks to classify parts automatically and uses the output layer neurons to directly represent part families. It extends the competitive learning model to overcome two of its biggest drawbacks. First, the parts are not always assigned a part family independently of the degree of similarity. ART neural networks use parameter called vigilance threshold ρ to ensure that similarity within a part family is not less than ρ, based on some similarity measure. Furthermore, the number of part families created does not have to be known a priori, but is determined during the clustering process.

The network architecture is similar to the architecture of competitive learning networks. The number of neurons in the comparison layer thus equals the maximum number of part families expected. Associated with each output neuron j, there is a weight vector $W_j = (w_{j1}, w_{j2}, ..., w_{jm})$ and an exemplar vector $T_j = (t_{j1}t_{j2}, ..., t_{jm})$. The exemplar vector is a binary representation of the weight vector, thereby representing the characteristics of the corresponding part family. Furthermore, recurrent connections are introduced, so that an input layer neuron I is connected to an output neuron j via a feedforward and a feedback connection with connection weights w_{ji} and t_{ji}, respectively.

For a better readability of the algorithm described below, we define $\parallel X \parallel = \sum x_i$. Note that for a binary vector X, this is simply the number of '1's in vector. Furthermore, let the intersection of to vectors $X \cap Y$ denote the vector C, whose elements are obtained by applying the logical AND operator to the corresponding elements in X and Y, implying that $c_i = 1$ if $x_i = y_i = 1$, and 0 otherwise.

1. Initialization: Select vigilance parameter ρ in the range $[0, 1]$. Initialize weight vectors with entries $w_{ji} = 1/(1 + m)$ and exemplar vectors with entries $t_{ji} = 1$ for all i, j.
2. Present an input pattern $X = (x_1, x_2, ..., x_m)$. For each output neuron j, compute $net_j = W_j \bullet X$ (weighted sum of inputs).
3. Select output node j^* with the highest net_j value. The exemplar vector T_{j^*} of this neuron is fed back to the input layer. To ensure similarity is higher than the threshold, check if

$$\frac{\| X \cap T_{j^*} \|}{\| X \|} > \rho$$

4. If similarity check fails, disable node j^* temporarily for further competition (to avoid persistent selection) and return to step 3.
5. If the similarity check is successful, set output value for j^* to 1 and 0 for all other output neurons. Update exemplar vector T_{j^*} to $\| X \cap T_{j^*} \|$. Furthermore, update weight vector W_{j^*} according to the equation

$$w_{j^*i} = \frac{x_i \wedge t_i}{0.5 + \| X \wedge T_{j^*} \|}$$

6. Enable any nodes disabled in step 5. Repeat steps 2.-7. until the last input pattern has been presented.
7. Repeat the entire process until the network becomes stable, meaning that the weight vectors stabilize to fixed values with only small fluctuations.

According to this algorithm, the ART network determines the cluster that is most similar to the current input pattern. Then, the similarity between the input and the cluster is computed as the fraction of perfectly matching '1's in the input and the exemplar vector in relation to the total number of '1's in the input vector. If this similarity measure is above the vigilance threshold, the input is assigned the corresponding class and the cluster?s weight and exemplar vectors are updated to represent the new element. If the similarity check fails, the procedure is repeated with the output neuron with the next highest *net* value. If the input can not be assigned any class, a yet uncommitted output neuron is selected eventually to represent this new class and its weight- and exemplar vectors are updated accordingly.

In the steady state, the connection weights also provide valuable information for the subsequent machine cell formation. A strictly positive connection weight w_{ji} implies that part family j contains at least one part that requires an operation to be performed on machine i. Therefore, considering the connection weight matrix can be used to easily assign machines to part families and also detect bottleneck machines and overlapping parts.

The ART network model provides the possibility to control the similarity among parts within one part family via the vigilance threshold and also does not require the number of part families to be specified in advance. Nevertheless, it also suffers some drawbacks. The biggest problem related to the traditional ART model is the *category proliferation problem* [Kaparthi et al. 1992]. During the clustering process, a contraction of the exemplar vectors can be observed, due to repeatedly forming the intersection with newly assigned input patterns. This leads to sparse exemplar vectors, which causes the network to create new categories frequently, as the number of unsuccessful similarity checks increases. Kaparthi and Suresh [Kaparthi et al. 1992] found out that because of this problem clustering is more precise when the density of the part-machine incidence matrices is high. As density is usually low, they propose to inverse the matrices for clustering purpose. This approach was further investigated and the improvements it brings to performance were confirmed by Kaparthi et. al. [Kaparthi et al. 1993]. Dagli and Huggahalli [Dagli et al. 1995] propose not to store the intersection of input- and exemplar vector, but the one that has the higher number of '1's. Chen and Cheng [Chen et al. 1995] state that the above methods may lead to improper clustering results and develop advanced preprocessing techniques to prevent category proliferation.

Another shortcoming is that clustering with ART networks is sensitive to the order in which the inputs are presented. Dagli and Huggahalli [Dagli et al. 1995] propose to preprocess the input vectors and present them in order of descending number of '1's, which also would help to address category proliferation, as the exemplar vectors would initially be taken from the inputs with dense feature vectors. Apart from the order of the inputs, clustering is very sensitive to the choice of the vigilance parameter. While it can be seen as an advantage to have the ability to control the size and number of clusters formed, it is also difficult to find the value that yields the best results.

2.4 Fuzzy Adaptive Resonance Theory

The Fuzzy ART model is the most recent development of neural networks for part-machine grouping. It tries to improve ART neural networks by incorporating the concepts of fuzzy logic. This model introduced by Carpenter, Grossberg and Rosen [Carpenter et al. 1991] and has also been applied to the cell formation problem in several works. For example, Suresh and Kaparthi [Suresh et al. 1994] use a Fuzzy ART network to form part families based on part-incidence matrices and compare their performance to matrix manipulation algorithms like ROC [King 1980] and also traditional ART, showing that Fuzzy ART networks are superior to these methods.

A fundamental difference of fuzzy ART networks in comparison with ART is the fact that they can handle both binary and non-binary input data. In order to incorporate fuzzy concepts into ART networks, occurrences of the intersection of two vectors applying a logical AND on the corresponding elements are replaced by the fuzzy AND operator $\wedge$, defined as $(x \wedge y) = min(x, y)$ [Carpenter et al. 1991]. In essence, the algorithm operates just as the ART network, with the following differences:

- The input of each output neuron j is computed as

$$net_j = \frac{\parallel X \wedge W_j \parallel}{\alpha + \parallel W_j \parallel}$$

- When checking for similarity, the fuzzy AND operator is used:

$$\frac{\parallel X \wedge W_{j^*} \parallel}{\parallel X \parallel} > \rho$$

- The weight update changes to incorporates the learning rate:

$$W_{j^*}^{new} = \beta(X \wedge W_{j^*}^{old}) + (1 - \beta)W_{j^*}^{old}$$

With the use of the fuzzy operator, the exemplars of the clusters are not restricted to binary values, and there is no need for a binary exemplar vector. Instead, both the bottom-up and the top-down connection have a weight denoted with $w_j i$, thus the weight- and exemplar vector of each output neuron are identical. In addition to the vigilance threshold, two more parameters are introduced, viz. the choice parameter α and the learning rate β. The learning rate specifies the speed with which the exemplars are updated in response to new inputs, which can be used to control the speed of learning and thereby reduce category proliferation.

Suresh and Kaparthi [Suresh et al. 1994] show that for the cell formation problem, applications using Fuzzy ART networks perform better and yield more consistent clustering results. This is especially due to the generalized update rule that can be controlled via the learning rate parameter. During the update process in ART networks, an entry was set to zero if the newly assigned pattern did not have this feature, which lead to rapid contraction of the exemplar vectors and thereby to category proliferation. In Fuzzy ART, the learning rate controls to what extend the exemplar entries are adjusted. For example, given a learning rate of $\alpha = 0.1$, an exemplar entry $w_{ji} = 1$ is adjusted to 0.9 the first time an input pattern without the corresponding feature is assigned to that cluster. In ART models, it would have been set to zero, reducing the probability of a pattern with that feature being assigned to that class significantly, and causing category proliferation in the long term. Note that a Fuzzy ART model with a learning rate of one and binary input vectors operates the same way as an traditional ART network.

The learning rate can also be adjusted to take advantage of another optional feature of Fuzzy ART networks, called *fast-commit slow-recode* [Carpenter et al. 1991]. This option involves a dynamic change to the learning rate that distinguishes between the first assignment of a pattern to a yet uncommitted node and the assignment to an existing cluster. In the first occurrence of a pattern, the learning rate is set to one to make the exemplar of the new cluster match the input vector. For subsequent assignments to that cluster, the updating process is dampened by choosing $\alpha < 1$. This method can help to incorporate rare features quickly in a new vector and prevent the quick deletion of learned pattern because of partial or noisy input data.

The Fuzzy ART model can be used for part family and cell formation in the same way as the ART model. But it provides several advantages over traditional ART, like the ability to handle continuous inputs, although the simple cell formation methods use binary part-incidence matrices as their input and thus do not yet take advantage of this ability. Furthermore, Fuzzy ART models provide more means to address the category proliferation problem, which is also critical in the context of group technology, as it is desirable to control the number of part families or machine cells created. The improved clustering precision is also due to the adjustable weight updates that prevent fast contraction of exemplar vectors and thereby increase the probability that similar input patterns will be grouped together, even when the corresponding vector has already been updated in between.

Because of their promising results, Fuzzy ART networks are subject of current research effort. The work by Suresh and Park [Suresh et al. 2003] introduces an extension to Fuzzy ART networks that allows them to consider part operation sequences during the clustering process. Peker and Kara [Peker et al. 2004] recently investigated on parameter setting for Fuzzy ART networks.

3 Directions for Future Work

Although neural network algorithms have been improved constantly to solve the clustering problem more efficiently, little effort has been made to solve the problem of bottleneck machines and overlapping. The majority of the methods implicitly duplicates machines and accepts overlapping parts that appear. Neural networks alone can not decide if it is advisable to duplicate a machine, because there is a more complex cost benefit analysis involved that requires information not available in the network. But due to the fact that many of the methods presented can be used to detect bottleneck machines, they may be integrated into decision support systems to help a decision maker or even be combined with other intelligent methods (e.g. expert systems) that are able to perform a cost benefit analysis.

Considerable research effort has been made to solve the unconstrained part-machine grouping problem. Currently, Fuzzy ART appears to be the most promising approach. Nevertheless, it currently uses fuzzy concepts only within the network and not yet at the input or output level, which could be a promising area for further research [19]. For instance, the network could indicate the degree of membership to each family with continuous values.

Although the neural network models have become more sophisticated over the years, there are still several challenges that have not received sufficient attention in literature so far. The methods described in this work focused on the unconstrained part and / or machine grouping based on the part's processing information or part features. In real world applications, several additional constraints have to be considered when designing a cellular factory layout. For instance, it may be necessary to consider part demands, capacity constraints or load balancing. Thanks to the development of more sophisticated clustering methods like Fuzzy ART networks and improved computational possibilities, these problems can be addressed in current and future research.

Rao and Gu [Rao et al. 1995] point out that the assumptions of unlimited capacity can lead to nonimplementable solutions and suggest an extended unsupervised network similar to the ART model to handle machine availability and capacities as additional constraints. This is mainly done by introducing a hidden layer with nodes representing each machine type that contain knowledge about the number of machines available and their capacity in the form of threshold functions. Parts can only be assigned to a part family if it does not violate any of these constraints.

Clearly neural networks will not be able to handle all constraints alone, but must be integrated with other methods. Some research effort has already been made in this area. For example, the work of Suresh and Slomp [Suresh et al. 2001] proposes a framework to include these additional constraints into the cell formation process. Theiy use a Fuzzy ART network first to form part families, before a mathematical goal programming model taking into account the additional constraints is used to form the corresponding manufacturing cells. Another interesting aspect of this work is the proposal to integrate this framework into a decision support system. As the execution time of the clustering process is fairly low by now, it is possible to have human interaction. This could help to make the complex decisions that require knowledge not available in the system, like considering the acquisition of new machines opposed to accepting inter-cell movements. Therefore, it is likely that future research will focus on extending the use of fuzzy concepts, integration of supplemental methods to consider more complex constraints and integration into decision support systems.

4 Conclusions

In this work, we gave an overview of neural network based methods proposed to address the problems related with the introduction of a cellular manufacturing layout. Neural networks have shown to be a promising tool for such complex clustering tasks with a great number of input elements, because of their greater flexibility and ability to learn from experience and generalize their acquired knowledge to recognize new input patterns. Current Fuzzy ART network provide consistent clustering results and an extension of the use of fuzzy concepts, as well as the integration of neural networks with other approaches appear to be promising research areas for the future.

References

[Burbridge 1991] Burbridge, J.L. (1991) *Production flow analysis for planning group technology.* Journal of Operations Management, 10(1), 5-27.

[Carpenter et al. 1991] Carpenter, G.A.; Grossberg, S.; and D.B. Rosen (1991) *Fuzzy ART: Fast stable learning and categorization of analog patterns by an adaptive resonance system.* Neural networks, 4, 759-771.

[Chen et al. 1995] Chen, S.-J.; and C.-S. Cheng (1995) *A neural network-based cell formation algorithm in cellular manufacturing.* International Journal of Production Research, 33(2), 293-318.

[Chu 1993] Chu, C.H. (1993) *Manufacturing cell formation by competetive learning.* International Journal of Production Research, 31(4), 829-843.

[Dagli et al. 1995] Dagli, C.; and R. Huggahalli (1995) *Machine-part family formation with the adaptive resonance theory paradigm.* International Journal of Production Research, 33(4), 893-913.

[Kao et al. 1991] Kao, Y.B.; and A. Moon (1991) *A unified group technology implementation using the backpropagation learning rule of neural networks.* Computers & Industrial Engineering, 20(4), 435-437.

[Kaparthi et al. 1992] Kaparthi, S.; and N.C. Suresh (1992) *Machine-component cell formation in group technology-a neural network approach.* International Journal of Production Research, 30(6), 1353-1368.

[Kaparthi et al. 1993] Kaparthi, S.; Suresh, N.C.; and R.P. Cerveny (1993) *An improved neural network leader algorithm for part-machine grouping in group technology.* European Journal of Operational Research, 69(3), 342-356.

[Kiang et al. 1995] Kiang, M.Y.; Kulkarni, U.R.; and K.Y. Tamb (1995) *Self-organizing map network as an interactive clustering tool - An application to group technology.* Decision Support Systems, 15(4), 351-374.

[King 1980] King, J.R. (1980) *Machine-component grouping in production flow analysis: an approach using a rank order clustering algorithm.* International Journal of Production Research, 18(2), 213-232.

[King et al. 1982] King, J.R.; and V. Nakorncha (1982) *Machine-component group formation in group technology-review and extension.* International Journal of Production Research, 20(2), 117-133.

[Kusiak 1987] Kusiak, A. (1987) *The generalized group technology concept.* International Journal of Production Research, 25(4), 561-569.

[Kusiak 1988] Kusiak, A. (1988) *EXGT-S: A knowledge-based system for group technology.* International Journal of Production Research, 26(5), 887-904.

[Peker et al. 2004] Peker, A.; and Y. Kara (2004) *Parameter setting of the Fuzzy ART neural network to part?machine cell formation problem.* International Journal of Production Research, 42(6), 1257?1278.

[Rajagopalan et al. 1975] Rajagopalan, R.; and J.L. Batra (1975) *Design of cellular production systems: A graph theoretical approach.* International Journal of Production Research, 13(6), 567-573.

[Rao et al. 1995] Rao, H.A.; and P. Gu (1995) *A multi constraint neural network for the pragmatic design of cellular manufacturing systems.* International Journal of Production Research, 33(4), 1049-1070.

[Suresh et al. 1994] Suresh, N.C.; and S. Kaparthi (1994) *Performance of Fuzzy ART neural network for group technology cell formation.* International Journal of Production Research, 32(7), 1693-1713.

[Suresh et al. 2001] Suresh, N.C.; and J. Slomp (2001) *A multi-objective procedure for labour assignments and grouping in capacitated cell formation problems.* International Journal of Production Research, 39(18), 4103-4131.

[Suresh et al. 2003] Suresh, N.C.; and S. Park (2003) *Performance of Fuzzy ART neural network and hierarchical clustering for part?machine grouping based on operation sequences.* International Journal of Production Research, 41(14), 3185-3216.

[Venugopal et al. 1992] Venugopal, V.; and T.T. Narendran (1992) *Neural network model for design retrieval in manufacturing systems.* Computers in Industry, 20, 11-23.

[Venugopal et al. 1994] Venugopal, V.; and T.T. Narendran (1994) *Machine-cell formation through neural network models.* International Journal of Production Research, 32(9), 2105-2116.

[Venugopal 1998] Venugopal, V. (1998) *Artificial neural networks and fuzzy models: New tools for part-machine grouping.* In: N.C. Suresh & J.M. Kay (Eds.): Group Technology and Cellular Manufacturing: State-of-the-art Synthesis of Research and Practice, Kluwer Academic Publishers, Norvell, Mass.

Part III

Socio-environmental complex modelling and territorial intelligence

The Evolution Process of Geographical Database within Self-Organized Topological Propagation Area

Hakima Kadri-Dahmani[1], Cyrille Bertelle[2], Gérard H.E. Duchamp[1], and Aomar Osmani[1]

[1] LIPN - University of Paris 13
99 avenue Jean-Baptiste Clément
93430 Villetaneuse, France
hkd@lipn.univ-paris13.fr,ghed@lipn.univ-paris13.fr,
ao@lipn.univ-paris13.fr

[2] LITIS, University of Le Havre
25 rue Philippe Lebon, BP 540
76058 Le Havre cedex, France
cyrille.bertelle@univ-lehavre.fr

Summary. The paper deals with Geographical Data Base evolution which is a major aspect of today's development in Geographical Information System (GIS). In a more practical aspect, GIS has now to evolve to manage updating. We will explain how the updating processes can be described as an evolution processus for GIS and transform them from complicated systems to complex systems.

Key words: geographical database; complex systems; evolution; emergence.

1 Introduction: GIS and their Evolution Process

A Geographic Information System (GIS) is a computer-based tool using geographical objects. A GIS is composed of a Geographical Data Base (GDB) with applicative operators which allow it to get, to stock, to verify, to manipulate, to analyze and to represent the spatial data of the GDB.

The originality of Geographical Data Base from ordinary Data Base is the use of spacial data [Rigaux et al. 2002]. The latter may be represented in a Geographical Data Base with two aspects: in raster mode or in vector mode. The raster mode is based on a pixel grid representation. The vector mode manipulates geographical features. In the following, we will focus our attention on the vector mode representation where each feature is represented by an object. Each object has a semantic part describing the nature of a feature which it represents and a geometric part describing its shape and its localisation. The position of the objects the ones compared with the others is an important information which is usually represented in the Geographical Data Base.

So, in Geographical Data Bases, geographical information is often represented with three levels: geometric, semantic and topological. From each level, we can define relations between objects that have to be linked corresponding to the specific level.

At the semantic level, a Geographical Data Base is often structured with layers. Layers are generally defined concerning a specific thematic like road traffic, fluvial tracing, building or vegetation. Generally, objects of a same layer have the same geometric representation and share the same topological properties inside networks.

The constant evolution of the real world which must be represented in the geographical data base induces the need of regularly update of the GDB and so make it evolve [Kadri-Dahmani et al. 2002]. To develop automatic evolution processes of a GDB, we must introduce inside the GDB itself, an adapted data representation containing relation between the objects.

In this paper, we present how GIS undergoing on evolution process, can be shown as a complex system. In section (3) we explain how the objects of the geographical data base interact at the time of an update. A crucial problem emerges then: which are the objects of the base concerned? We propose in section (4) a propagation algorithm which allows the propagation of interactions. From this algorithm emerges a property given in the section (5). Section (6) presents implementations and experiments and we conclude in section (7).

2 From GIS to Complex GIS

All these structured informations which defines a GIS introduce a great number of static dependences but each layer can be generally understood alone or some parts of each layer can be isolated to better understand the dependence between involved objects. Generally the applicative operators can be computed on each of these parts. In that way, we can consider classical GIS as complicated systems in the terminology proposed by Le Moigne [Le Moigne 1999]. We can consider that the Geographical Data Base in association with the previous applicative operators which constitute the GIS is a closed system.

Today, the complexity of the world requires to use or to add additional functionalities on GIS. Geographical informations deal also with human-landscape interactions. The simulation of social aspects and of ecological processes seems to be more and more linked to the better understanding of the geographical data and its evolution inside its all social, geopolitic and ecological environment. To integrate these new aspects, we have to manage some complex processes like some energetic fluxes that crosses the standard GIS (see the figure 1). This complex fluxes transform the standard GIS in open systems which confer to them some properties linked to

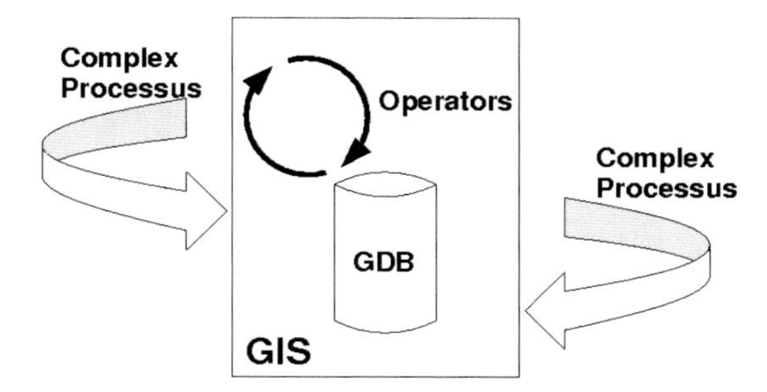

Fig. 1. GIS under Complex Processus (uodating flow)

complexity. Self-organization and multi-scale organizations can emerge from these complex processes. The expected evolutions of GIS can be considered as the transition which will transform the standard complicated GIS into complex GIS. We say that GIS is under an updating flow.

In the following, we deal with a specific improvement on GIS which concerns its own evolution. As described in the previous section, the constant evolution of the real world induces the need of regularly updating the geographical data of GIS. This evolution processus is typically a complex processus that generates some dynamical organizational processes inside GIS. The data themselves retro-act on the processus during the propagation method that we will present in the next sections [Kadri-Dahmani 2005].

3 Evolutive GIS Formalism

We adopt the feature-based approach, where features are the fundamental concept for the representation of geographical phenomena as described in [Kadri-Dahmani 2005]. Basically, a GDB is represented in a minimal formalism, by the pair (V, D) where:

1. V is the set of the classes used in the GDB. Each class gathers features which have common characteristics. The set V gathers GDB scheme elements.
2. D is the definition domain of the variables of V. It is the set of the objects of one GDB instance. The heterogeneity of objects which belong to a GDB needs their classification for a better use. We consider 4 classes of objects which are spread over two information levels: the geometric level which gathers geometric primitives PG and the geometric objects OG, and and the semantic level which gathers simple semantic objects OS and complex semantic objects OC.

$$D = PG \cup OG \cup OS \cup OC$$

The proposed model for a GDB which allows to evolve through updating operations must add some complementary sets which will manage some dependences between the geographical elements. So the model will be composed of a quadruplet (V, D, R, C) where, in addition:

3. The connection graph over the GDB elements is based on relations between these elements. R is the set of these relations. The different kinds of relations that we consider are: composition relations RC, dependence relations RD and topologic relations RT

$$R = RC \cup RD \cup RT$$

   In this paper, we focus our study on topological relationships. Topological relations are fundamental and allow to describe the relative position of the objects with each others. There is many models which propose a representation of this kind of relation, we propose to use the 9-intersection model from Egenhofer and Herring [Egenhofer et al. 1991]. In this model, each object p_i is defined by the inset, noted $\dot{p}_i$, the outline set, noted δp_i, and the exterior set, noted $\bar{p}_i$. This model can be represented by a matrix formulation. A topological relation between two objects X and Y is represented by a matrix. Each element of this matrix denotes the intersection between different components of X and Y [Egenhofer et al. 1991].
4. C is a set of constraints defined between the variables of V and/or between variables value of V. In our object modelization, this corresponds to constraints defined between the classes (constraints between variable) and/or between objects (constraints between values). These constraints manage the GDB evolution on many levels. This quadruplet corresponds to the GDB modelisation to prepare it to evolution.
5. Finally, to effectively manage evolution processes, we have to modelize the updating informations in accordance with the GDB conceptual model. We note M the updating set where basis action is the transaction and the full model for the GDB is the 5-uplet (V, D, R, C, M).

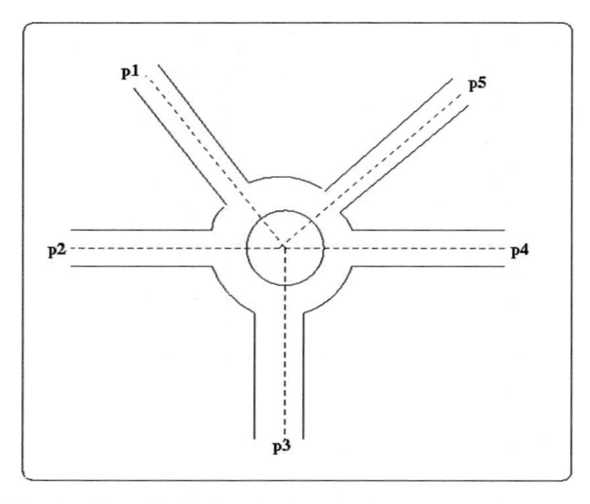

Fig. 2. Round-about object geometric representation

Example 1. Figure 2 shows a geometric representation of a Round-about object O_i. O_i is a complex object composed of 5 simple objects. Each two simple objects are connected by a topological relation "touch".

4 Updating Propagation: Dynamical Interaction Network

The integration of the updating flow in the geographical data base starts the dynamics of the system evolution. Let's recall that updating flow is structured as a set of transactions, each one is a set of canonical operations sequences. So, each canonical operation, applied on an object, activates some falls of updating operations to apply on other objects. At the topological level, geographical objects are linked with topological relations. The evolution of an object X in topological relation with the object Y, needs also the evolution of this last one. The evolution of Y will need also to make evolve the other objects in relation with it. Usually the evolution of X, called starting operation, is known since it is a membership of the input updating flow, but its influence on the objects linked to X is not known. So, consequent operations, called influence operations, are not known. Objects to which these operations are applied, called influenced objects, are not either known.

So, propagate the effect of an updating operation in the geographical data base means to execute the starting operation and all the influence operations which result from it without altering its consistency. This is translated, in our system, by the installation of an interaction network built from the connection graph and the table of influences in [Kadri-Dahmani 2005].

ID	CLASS	RELATION
7130	BATIMQCQ	contains
7135	BATIMQCQ	contains
7134	BATIMQCQ	contains
3640	ROUTE TR	borders
2264	ROUTE TR	borders
4520	CARREFOURNA	touchs
4530	CARREFOURNA	touchs
...	...	...

Table 1. Topological influence area from object ID 9814

If we consider an object O_c and an associated updating operation, *op*, the mecanism of propagation is applied in a local zone centered on the object O_c, called working zone from where we extract the set of other objects which may be under the influence of the first one. The propagating mechanism is recursif but we limit the exploration inside the working zone. The step is presented in the Propagation algorithm (see Fig. 5) where InfluenceAreaT() allow to delimit the Topological Influence Area defined as follow:

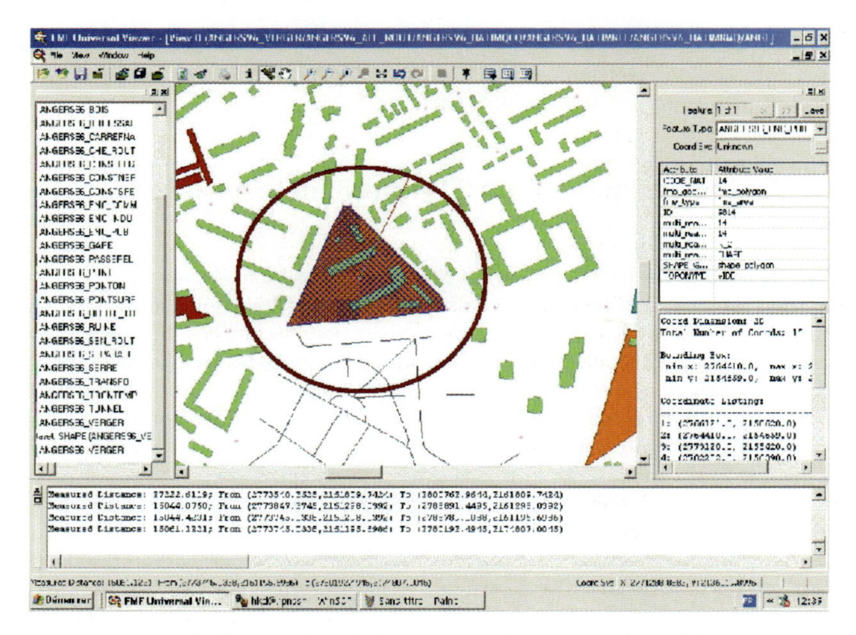

Fig. 3. Topological influence area from object ID 9814

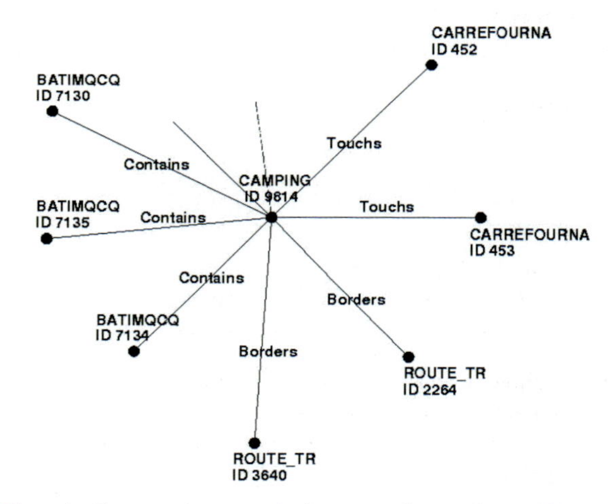

Fig. 4. Connection graph (extract from object ID 9814)

```
Propagate(Op1, O1, O2, Rel, tabu, Result, Zte,BC):Result
01 - if Op1=Identity or O2=Null or Rel=Null
02 - then
03 -    if O2≠Null
04 -    then tabu <- tabu - {O2}
05 -    else tabu <- tabu - {O1}
06 -    endif
07 -    Result <- true
08 - else
09 -    Op2 <- InfluenceTableVisit (Op1, O1, O2, Rel)
10 -    Zitc(O2) <- InfluenceAreaT(O2, Zte).Zitc(O2)
11 -    if DirectUpdate (Op2, O2, Zitc(O2),BC).Success
12 -    then
13 -       Result <- Result ∪{O2}
14 -       Success <- true
15 -       For all object O3 ∈ Zitc(O2) then
16 -          if (O3 ∉ tabu)
17 -          then
18 -             Rel2 <- Relation(O2,O3)
19 -             Success <- Propagate (Op2, O2, O3, Rel2,
                                      tabu ∪{O3}, Result, Zte, BC)
20 -          endif
21 -       endfor
22 -    else Success <- False
23 -    endif
24 - endif
25 - return success
```

Fig. 5. Propagation Algorithm

Definition 1. *We consider O_c a target object to be updated from a GDB. Its Topological Influence area that we note as $Zit(O_s)$, collects all the GDB objects which are inside a working zone of O_c and which are linked to O_c.*

Example 2. Figure 3 shows working zone of the target object ID 9814 which is a camping. The table1 contains a part of the set of the ID objects which belong to his influence area. For each ID object, its corresponding class and the topological relation with the target object are given.

We represent the topological influence area by a connection graph which connects the objects linked by a topological relation.

Example 3. Figure 4 shows some elements of the connection graph corresponding to some relation from the target object ID 9814, based on the table 1.

5 Emergent Property

The local propagation allows to avoid to explore the whole geographic data base. Now, we need to build a processus that will compute the adapted working zone which permits to tell whether the local consistency maintenance is enough to insure the global consistency maintenance. For that purpose we propose an algorithm that we called dilatation method and that consist to progressively increase the working area like a new disk centered on the initial object O_c and which radius is augmented step by step with the value p until that a further increase will not compute new objects involved in the propagation processus. That leads us to define this computed area as the stability area associated to the object Oc. To prove that the local consistency maintenance can be sufficient for the global consistency maintenance, we had to define some hypothesis about the regularity of the objects repartition. The properties given in the following are prove in [Kadri-Dahmani 2005].

Definition 2. *A finite set of planar points is called* **p-dense** *if the Delaunay triangulation over all the set of points has no edges longer than p.*

Property 1. If the influence area of the point O_c is p-dense then the dilatation method with a step equal to p computed from O_c give the stability area of this point.

Property 2. The local consistency over the stability area for an object O_c will insure the consistency of the whole Data Base.

This last emergent property allows us to define a subset of objects from the GDB and be able to predict that the behavior of these objects is himself the behavior of all objects of the GDB vis-a-vis to a flow of update. This first theoretical result allows us to implement in an efficient way, the whole updating system, with the postulate that the natural geographical data follows this hypothesis of regular distribution, using an adapted step of resolution for the dilatation method.

6 Implementation and Experiments

The whole system has been originally developed in the COGIT laboratory where it has been implemented. This system is in operational practice and has been connected to the framework OXYGENE [Badard et al. 2003] of this laboratory. OXYGENE, mainly developed in Java, allows to model and to use geographical information within oriented object scheme. Data Bases are managed with Oracle which has been completed with a spacial data extension. The mapping between Oracle and Object representation is made by JDO (Java Data Object) and OJB.

A methodology for its validation has been developed and has proved that the mechanisms are efficient, even if some rejection can be avoid with a better scheduling. An experiment has been developed on the IGN GDB concerning the Angers French town zone (see Fig. 6).
Using some matching technique between two of these GDB from 1994 and 1996 (respectively called BDTopo94 and BDTopo96), we have built a set of updating informations. We define 2 indicators to validate the automatic updating process respecting the consistency maintenance:

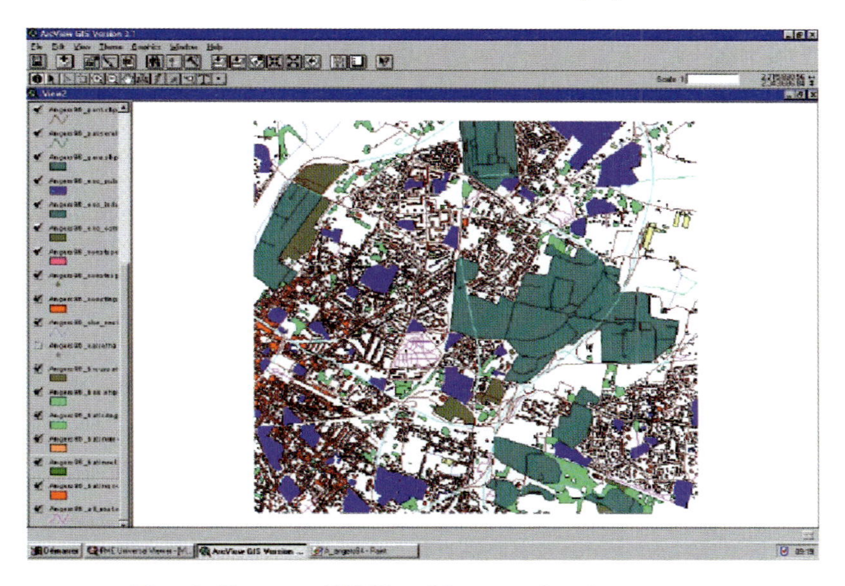

Fig. 6. Extract of BDTopo96 concerning Angers town

- The precision indicator defined as the rate of correct updating decision number over the total realized decision number;
- The similarity indicator which compares the updating results from BDTopo94 with the true situation in BDTopo96. This indicator is the rate of the number of matched objets in this comparison over the total number of matched objects and unmatched objects.

From a specific experiment based on 30 canonical operations, we have obtained a precision indicator equal to 0.948 and a similarity indicator equal to 0.962 [Kadri-Dahmani 2005] which is sufficient to validate the whole processus.

7 Conclusion

This paper describes a consistent updating processus over a Geographical Data Base (GDB) as a complex operation concerning Geographic Information Systems (GIS). Our purpose is to explain where the complexity occurs during the processus, The formalism proposed for the geographical information description is based on objects. To manage the updating, we have to define semantic and topological relations which allow to define the influence area associated with each object. These relations between objects are represented with a connection graph.

Moreover, we have to manage some constraints which deal with consistency maintenance of the whole data base ; a specific language is proposed for that purpose. Even if the connection graph is important on a whole geographic map, the previous system to describe GIS is complicated in the sense that we can manage it correctly by successive splitting and application of basic operators.

The updating processus is then defined over the GIS as a complex flux that make involved the GIS. This processus implements a propagation method which consists to act by updating on GDB objects and these objects propagate the updating operators using an influence table and so retro-acts on the whole system and processus. In that sense, the updating processus crosses the GIS like an evolutive organizational flux which transform the GIS from a complicated system to a complex system. The basis of the updating is the use of influences tables which summarize all the canonical operators needed. The application of this influence table can be compared with the rule based processus which make involved a cellular automaton.

Finally we show how the updating processus, as a complex flux over GIS, can lead to obtain an emergent property. This property allows to obtain the global consistency maintenance of the whole GDB from only local consistency maintenance. We implement a dilatation method that can be considered as a way to obtain a self-organization concerning the updating problem.

The complex decomposition and description of the work presented in this paper allows us to build conceptual models over GIS which can be used to manage some others kinds of complex processes. We can adapt this proposed method for updating flux to other kinds of complex processes flux. In these complex processes, we can consider the human aspects of geography which deal with social, geopolitic and ecological purposes (6). The proposed methods used here can give conceptual approaches to manage such major developments which give all the power in the use of GIS in our present complex world, to better understand and analyze it.

References

[Badard et al. 2003] Badard, T.; and A. Braun (2003) *OXYGENE: an open framework for the deployment of geographic web services.* in Proceedings 21th International Cartographic Conference, Durban, South Africa.

[Egenhofer et al. 1991] Egenhofer, M.; and J.R. Herring (1991) *Categorizing Binary Topological Relations Between Regions, Lines and Points in Geographic Databases.* Technical Report, Departement of Surveying, University of Maine.

[Kadri-Dahmani 2001] Kadri-Dahmani, H. (2001) *Updating in GIS: Towards a more generic approach.* in Proceedings 20th International Cartographic Conference, Beijing, China.

[Kadri-Dahmani et al. 2002] Kadri-Dahmani, H.; and A. Osmani (2002) *Updating Data in GIS: How to maintain Database Consistency?.* in Proceeding 4th International Conference of Entreprise Information system, Ciudad Real.

[Kadri-Dahmani 2005] Kadri-Dahmani, H. (2005) *Mise à jour des Bases de données géographiques et maintien de leur cohérence.* Ph.D. Thesis, University of Paris 13.

[Le Moigne 1999] Le Moigne, J.-L. (1999) *La modélisation des systèmes complexes.* Dunod.

[Rigaux et al. 2002] Rigaux, P.; Scholl, M.; and A. Voisard (2002) *Spatial Databases - with applications to GIS.* Morgan Kauffmann.

Self-Organization Simulation over Geographical Information Systems Based on Multi-Agent Platform

Rawan Ghnemat[1], Cyrille Bertelle[1], and Gérard H.E. Duchamp[2]

[1] LITIS, University of Le Havre
25 rue Philippe Lebon, BP 540
76058 Le Havre cedex, France
rawan.ghnemat@gmail.com, cyrille.bertelle@gmail.com

[2] LIPN - University of Paris 13
99 avenue Jean-Baptiste Clément
93430 Villetaneuse, France
ghed@lipn.univ-paris13.fr

Summary. In this paper, we present a review concerning the coupling of Geographical Information Systems with agent-based simulation. With the development of new technologies and huge geographical databases, the geographers now deal with complex interactive networks which describe the new Geopolitics and world-wide Economy. The aim here, is to implement some self-organization processes that can emerge from these complex systems. We explain how we can today model such phenomena and how we can implement them in a practical way, using the concept of complex systems modelling and some efficient tools associated to this concept.

Key words: Complex systems, Self-organization, Geographical Information System (GIS), Individual-based model (IBM), Agent-based modeling and simulation (ABMS)

Acknowledgment

The authors have developped this paper to focuss on an innovative methodology to develop geographical, social and environmental system simulations using and highlighting many references from books as the one edited by H.R. Gimblett [Gimblett 2002], or from unpublished web-based documents [Gessler 2008, Repast 2008, Agent Analyst 2008, Ligman-Zielinska 2005]. Many of the illustrations used here, are made under the courtesy of these document authors.

1 Introduction

The challenge of the simulation of complex systems modelling based on Geographical Information System (GIS) is to propose the future decision making supports for urban plannings or environmental, social-politic development. Geographers manage today a great amount of geographical data and need innovative methodologies to analyse them in a pertinent way. A complex vision of the current world is strongly needed in order to face nowadays challenges in understanding, modelling and simulating [Aziz-Alaoui et al. 2006]. The increasing of intensive communications, allowed by high technologies, leads to develop in an efficient way, the information access and sharing. This revolution deeply transforms Geopolitics and world-wide Economy into geographical complex systems in huge interactive networks. Mixing GIS with complexity modelling is a new challenge to understand, analyse and build decision support systems for Economy and Geopolitics.

The actual GIS conceptual model is based on a layered structure [Goodchild et al. 1991]. A layer allows gathering, in a same set, some objects corresponding to a specific thematic, hydrological layer, building layer, road layer, ... Some new challenges for GIS future development, is to use the new conceptual models given by complexity theories to automatically identify some dynamical organizations and so to manage scenario of developments which can be used to determine automatically some social or urban re-organizations, for example.

The use of Individual-Based Model (IBM) offers potential for studying complex system behaviors and human/landscape interactions within a spatial framework. Artificial intelligent agents introduce behavior conditions and set communications or interactions between them and their environment as the major rule of the simulation evolution. Agents have goals who lead their interactions or actions over the world. Few researchers have mixed spatially explicit agents and GIS. After a non exhaustive review on that subject, we will focus our attention on Agent Analyst, based on Repast Multiagent system [Agent Analyst 2008,Repast 2008]. Agent Analyst can be presented as a free extension to ESRI's products like ArcGIS.

2 Mixing Individual-based Models and GIS

The use of Individual-Based Model (IBM) is a promising approach to model spatially explicit ecological phenomena. Interest has increased in using GIS for simulation to spatial dynamic processes. A great part of the challenge of modeling interactions between natural and social processes has to do with the fact that progress in these systems results in complex temporal-spatial behavior [Gimblett 2002].

Successive efficient computer science concepts used for that purpose, are:

- Cellular automata theory which has demonstrated efficiency about modeling landscape dynamics by regular rule-based processes;
- Object-oriented modeling (OOM) which has proved its power by representing the domain with concrete objects that have as much similarity with their real world counterparts as possible;

- The spatially explicit model mixed with object-based modeling lead to define the individual-based modeling (IBM). The characteristics and advantages of IBM are [Gimblett 2002]:
  - A variety of types of differences among individuals in the population can be accommodated;
  - Complex system decision making by an individual can be simulated;
  - Local interactions in space and the effects of stochastic temporal and spatial variability are easily handled.
- Artificial intelligent agents (Wooldridge, 2002) introduce behavioral conditions and set communications or interactions between them and their environment as the major rule of the simulation evolution. Agents have goals who lead their interactions or actions over their world.

Few researchers have mixed spatially explicit agents and GIS. The first goal for us, is to have at hand, a virtual laboratory in order to study the outcomes of various behavior on realistic landscapes.

Jiang and Gimblett provide an example of modeling pedestrian movements using virtual agents in urban spaces. Corresponding to cognitive processes for their behavior, agents act relatively to their own environment perception. Decision makers, such resources managers or designer, faced with realistic environment problems, would substantially benefit from these simulation techniques.

Itami and Gimblett have developed the Recreation Behavior Simulator (RBSim) to simulate the behavior of human individuals in natural environments. RBSim joins two computer technologies: GIS to represent environment and Autonomous Human Agents in order to simulate human behavior within geographic space. This tool can allow us to investigate tourism management options. It tries to give answers to some questions like: how different management options might affect the overall experience of tourists? How to schedule some visits and know the impact on the number and frequency of users? A graphical output of RBSim is illustrated in the figure 1.

Randy Gimblett develops studies about Colorado River through the Grand Canyon national park. This natural site suffers from impacts of repeated recreational uses like campsites and the destruction of sensitive vegetation due to many visitors. A management plan has been established to protect the national resource. RBSim is currently used to simulate the current use pattern of commercial tour operators in the Grand Canyon. In addition, managers wish to test alternative management strategies including alternative timetables for starting trips, booking specific riverside campsites for different sized boating parties: these strategies being aimed to reduce conflicts between different boating parties, reducing environmental impact on river beaches, and increasing the number of river users.

After this brief review, we will present in the next section, a self-organized spatial model which will be implemented in a mixed GIS (Geographical Information System) - ABMS (Agent Based Modeling and Simulation) model for urban development.

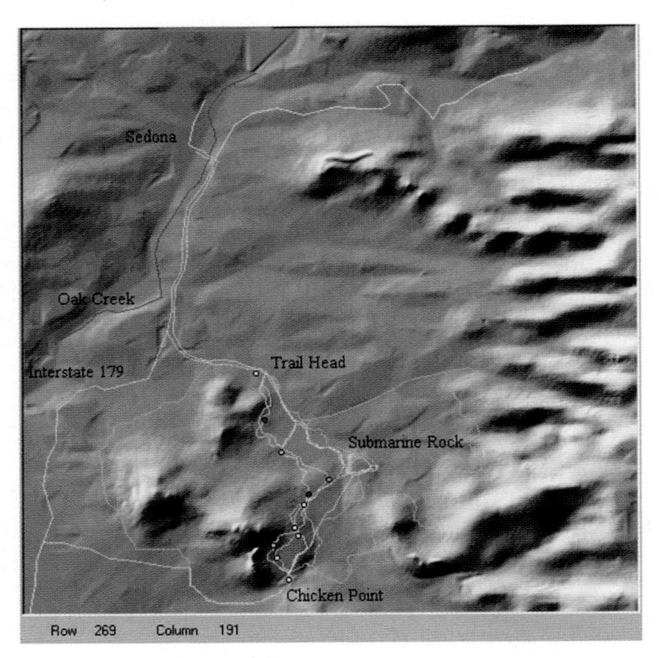

Fig. 1. Dynamical agents move across the landscape in RBSim [Gimblett 2002].

3 Schelling's Self-Organized Segregation Model

Thomas Schelling received in October 2005, the Nobel Price in Economic Sciences. He contributed to enhance the understanding of conflicts and cooperation about social institutions. His major work deals with game theory, he proposes a simple model of spatial segregation which can lead to self-organized phenomena. Schelling's city segregation model illustrates how spatial organizations can emerge from local rules, concerning the spatial distribution of people which belong to different classes. In this model, people can move, depending on their own satisfaction to have neighbours of their own class. Based on this model, a city can be highly segregated even if people have only a mild preference for living among people similar to them.

In this model, each person is an agent placed on a 2D grid (in his original presentation, a chessboard was used by Thomas Schelling). Each case can be considered like a house where the agent lives. Each agent cares about the class of his immediate neighbours who are the occupants of the abutting squares of the chessboard. Each agent has a maximum of eight possible neighbours. Each agent has a "happiness rule" determining whether he is happy or not at his current house location. If unhappy, he either seeks an open square where his happiness rule can be satisfied or he exits the city. The rule-based system is described as following:

- An agent with only one neighbour will try to move if the neighbour is of a class different than his;

- An agent with two neighbours will try to move unless at least one neighbour is of the same class as his;
- An agent with from three to five neighbours will try to move unless two neighbours are of the same class as his;
- An agent with from six to eight neighbours will try to move unless at least three neighbours are of the same class as his.

The exact degree of segregation which emerges in the city depends strongly on the specification of the agents happiness rules. It is noticeable that, under some rule specifications, Schelling's city can transit from a highly integrated state to a highly segregated state in response to a small local disturbance. We can observe some bifurcation phenomena which lead to chain reaction of displacements.

In the figure 2, we present some results of Schelling's model computed on a cellular automaton based on the applications proposed by N. Gessler [Gessler 2008].

4 Mixing Agent-Based Simulation and GIS Model Support to Implement Self-Organized Segregation Model

Agent Analyst is an Agent-Based Modeling and Simulation (ABMS) extension for the ESRI's ArcGIS suite of products [Agent Analyst 2008]. Agent Analyst integrates and extends the functionalities of the open-source Repast modeling and simulation. The Recursive Porous Agent Simulation Toolkit (Repast) is an agent modeling toolkit [Repast 2008]. It borrows many concepts from the Swarm agent-based modeling toolkit, but it multiple pure implementations in several languages and built-in adaptive features such as genetic algorithms and regression. Repast seeks to support the development of extremely flexible models of living social agents, but is not limited to modeling such entities.

Agent Analyst fully integrates ABMS with GIS. Through this integration, GIS experts gain the ability to model behaviours and processes as change and movement over time (e.g., simulate land use and land cover changes, predator-prey interactions, or network flows and congestion) while ABMS modellers are able to incorporate detailed real-world environmental data, perform complex spatial processes, and study how behaviour is constrained by space and geography. Furthermore, ABMS models can include real-time GIS data for situations such as disaster management, fire fighting, or resource management.

The graphical Agent Analyst tools allow the user to create agents, schedule simulations, establish mappings to ArcGIS layers, and specify the behaviour and interactions of the agents. In the following, we will show how using Agent Analyst for Schelling"s problem. A sample will be given using GIS and agent model and develop the use to implement it.

In the following, the Schelling's segregation model is applied on a GIS vector layer rather than a 2D grid [Ligman-Zielinska 2005]. The implementation is made using Agent Analyst over ArcGIS. The model is composed of two main parts. The first one is an environment layer of zip code regions from a GIS, represented in Agent

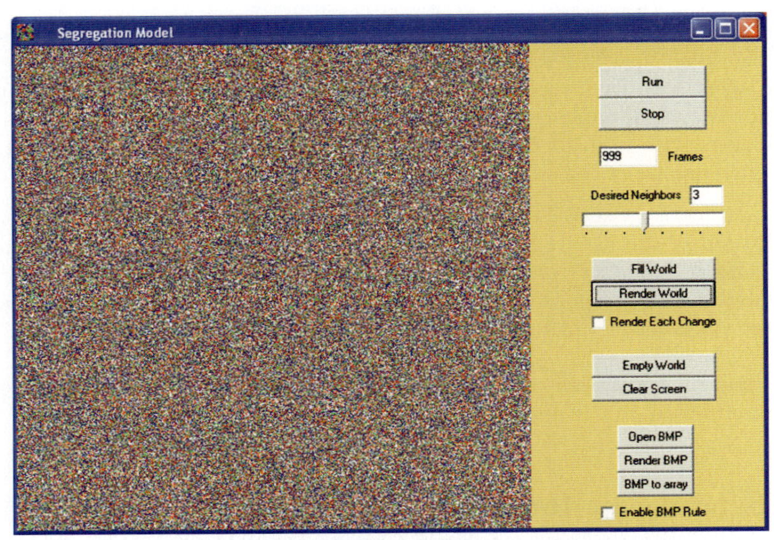

(i) Initial configuration

(ii) After 999 steps configuration

Fig. 2. Schelling's segregation model implemented by N. Gessler [Gessler 2008]

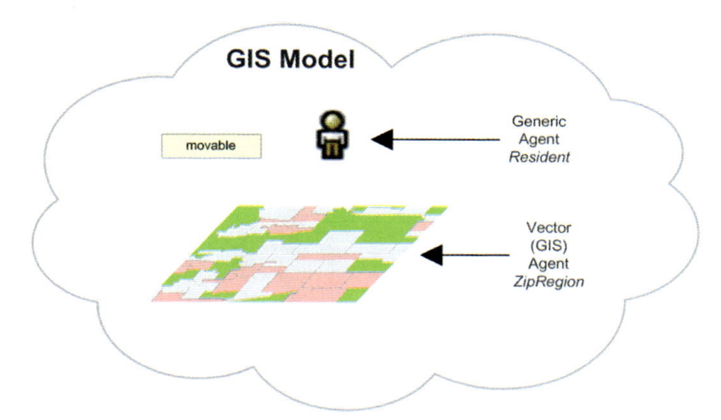

Fig. 3. Agent-GIS mixing after [Agent Analyst 2008]

Analyst by a non movable vector GIS agent. The other part is a set of movable generic agents, which represent the city residents who make decisions of moving to a new location (zip region). This general architecture is represented by figure 3.

The basis of the resident agent displacements come from the local rules of preference described below and a segregation phenomenon emerges from the interplay of these individual choices.

The model assumptions are based on a twofold distinction, red and green resident identity. The model is equipped with a basic parameter threshold called tolerance, which reflects agent preference level of neighboring agents that are alike (i.e. the same "color"). For example, if the tolerance parameter is set to 30%, then it is assumed that the agent will be satisfied if the ratio of identity-different neighboring agents to total number of neighbors does not exceed 30%. If it does, then the agent is dissatisfied and moves out to another region that is currently unoccupied.
Using Agent Analyst, we load the city layer shape in ArcMap as represent in figure 4.

Then the Schelling model is loaded in the ArcToolbox defined for AgentAnalyst as process. Figure 5 is the interface obtained. We can find on this interface a lot of elements like display name or GIS package used (ArcGis or OpenMap) but also Actions editor.

Actions editor was designed to automatize the process of programming model actions. It composes all the necessary elements that constitute an action. Since an action is a method in object-oriented programming language, which is here a variant of Python, we need to define its name, import the necessary classes (modules) from outside the model (if we need any), use pre-existing variables, and finally write the code for action.

Once all elements are defined, we can compile and run the model. One possible issue of Schelling's model in the initial map defined previously can be represented

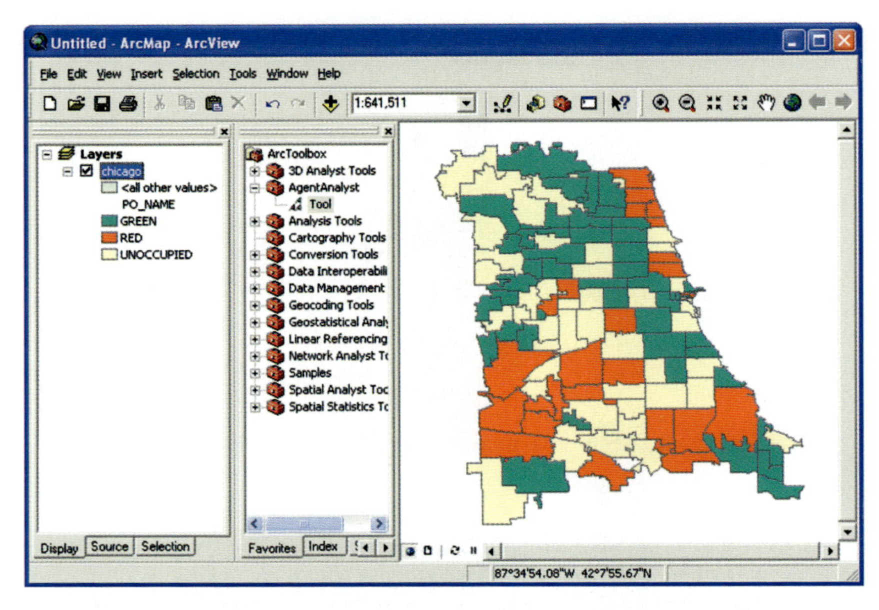

Fig. 4. City layer shapefile after [Ligman-Zielinska 2005]

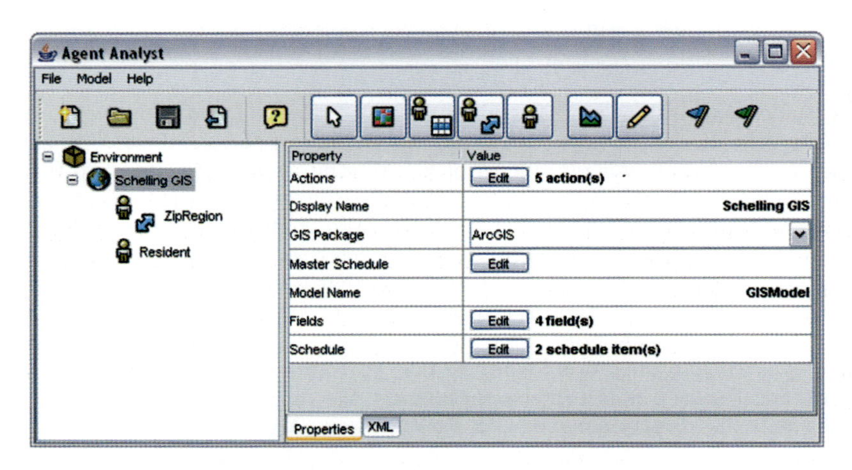

Fig. 5. Agent Analyst model component

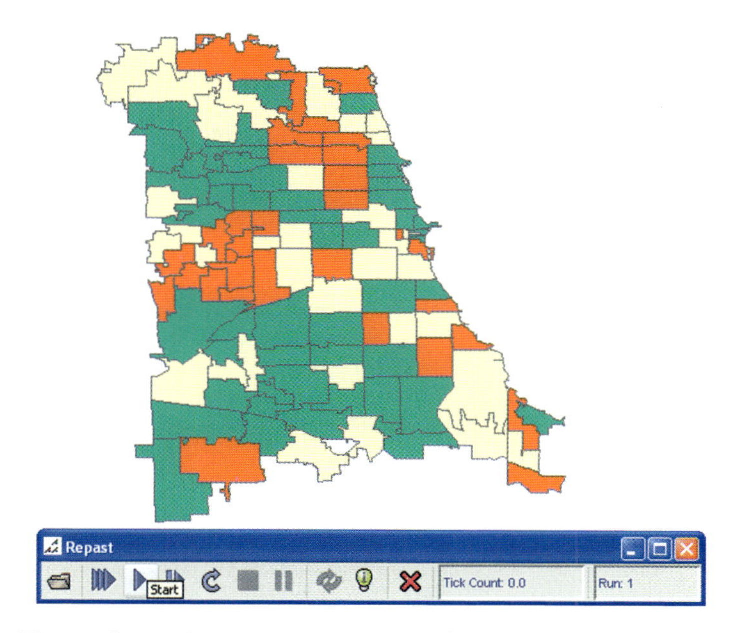

Fig. 6. Agent Analyst simulation after [Ligman-Zielinska 2005]

by figure 6. In this figure we have the Repast toolbar which allows managing the simulation execution.

Even if segregation is an important problem to manage or to control the urban systems development, the schelling model can be used and generalized to many others problems, especially for non structured networks. A. Singh and M. Haar [Singh et al. 2004] have proposed an extension of Schelling's model for managing peer to peer (P2P) networks. P2P is a new way of decentralized organization for communication over networks. Because of the increasing possibilities to share and to distribute efficient computers at low cost on many position on a network, the P2P way of communication could be one of the major in the future. The robustness of such networks in any kind of critical situations can make them powerful and efficient.

In an unstructured decentralized network, the location of the peers is decided randomly and therefore peers with high bandwidth may be adjacent to peers with low bandwidth introducing undesired low bottleneck bandwidth in the network. Schelling's method is based on local displacements, which change the peers connections of the considered computer over the whole connection graph. Schelling's method is also based on satisfaction criteria which is the fact that the neighbors have similar bandwidth than the considered computer. This demonstrate of Schelling's model can be used effectively for adapting P2P network topology and so to improve bandwidth usage. We can generalize this study to any kind of unstructured networks such as mobile telecom infrastructures distribution over an urban system area.

5 Conclusion and Perspectives

This paper presents some new methodologies using agent-based modelling to implement the complexity needed to simulate geographical, social and environmental systems. The first part shows a review of the usage of simple mixing of agent representation interacting on GIS. The second part shows how this mixing can allow to implement more complex situations, dealing with self-organization processes. In that purpose, we describe an example of such self-organization processes by the implementation of Schelling's segregation model over a Geographical Information System, using mixing with Agent Based Modeling and Simulation. Extension of this segregation model can be found to P2P network management but also more generally, to interaction network management. The mixed GIS-ABMS platform is intended now to be connected to ant systems [Bertelle et al. 2006] and automata models to allow to represent evolutive negociation processes [Ghnemat et al. 2006]. Applications to economy and urban development are well-suited to this forthcoming studies.

References

[Agent Analyst 2008] Agent Analyst web site (2008) `http://www.institute.redlands.edu/agentanalyst/AgentAnalyst.html`.

[Aziz-Alaoui et al. 2006] Aziz-Alaoui; M.A.; and C. Bertelle editors (2006) *Emergent Properties in Natural and Artificial Dynamical Systems.* Collection "Understanding Complex Systems", Springer.

[Bertelle et al. 2006] Bertelle, C.; Dutot, A.; Guinand, F.; and D. Olivier (2006) *Organization Detection using Emergent Computing.* Int. Transactions on Systems Science and Applications, Special Issue *Self-Organizing, Self-Managing Computing and Communications*, vol. 2(1), 61-69.

[Gessler 2008] Gessler, N. (2008) *Building Complex Artificial Worlds.* `http://gessler.bol.ucla.edu/`, `http://www.sscnet.ucla.edu/geog/gessler/borland/segregation.htm`.

[Ghnemat et al. 2006] Ghnemat, R.; Oqeili, S.; Bertelle, C.; and G.H.E. Duchamp (2006) *Automata-Based Adaptive Behavior for Economic Modeling Using Game Theory* in M.A. Aziz-Alaoui and C. Bertelle (eds) *Emergent Properties in Natural and Artificial Dynamic Systems*, Understanding Complex Systems series, Springer.

[Gimblett 2002] Gimblett, H.R editor (2002) *Integrating Geographic Information Systems and Agent-based Modeling Techniques.* Santa Fe Institute studies in the sciences of complexity, Oxford University Press.

[Goodchild et al. 1991] Goodchild, M.F.; Rhind, D.; and D.J. Maguire (1991) *Geographical Information Systems: Principles and Applications*, Longman, New York.

[Ligman-Zielinska 2005] Ligman-Zielinska, A. (2005) *Agent Analyst tutorial - Schelling GIS*, San Diego University.

[Repast 2008] Repast web site (2008) `http://repast.sourceforge.net`.

[Singh et al. 2004] Singh, A.; and M. Haar (2004) *Topology adaptation in P2P Networks using Schelling's model.* Workshop on Emergent Behavior and Distributed Computing, PPSN-VIII.

[Wooldridge 2002] Wooldridge, M. (2002) *An introduction to MultiAgent Systems*, John Wiley & Sons.

Cliff Collapse Hazards Spatio-Temporal Modelling through GIS: from Parameters Determination to Multi-scale Approach

Anne Duperret[1], Cyrille Bertelle[2], and Pierre Laville[3]

[1] LOMC FRE 3102, University of Le Havre
25 rue Philippe Lebon, BP 540
76058 Le Havre cedex, France
anne.duperret@univ-lehavre.fr

[2] LITIS, University of Le Havre
25 rue Philippe Lebon, BP 540
76058 Le Havre cedex, France
cyrille.bertelle@univ-lehavre.fr

[3] BRGM - Maison de la Géologie
77 rue Claude Bernard
75005 Paris, France
p.laville@brgm.fr

Summary. In this paper, we study the cliff collapses, using observations and in situ measures, along 120 km of the french chalk coastline in Upper-Normandy and Picardy. Cliff collapses occur inconsistently in time and space, in unpredictable way. A european scientific project ROCC (Risk Of Cliff Collapse) has been launched (1999-2002) in order to identify the critical parameters involved, to evaluate their impact and their interaction in mass movements. Cliff collapse process appears as a complex natural system, due to the large amount of parameters able to lead to a collapse. GIS approach has been used to allow an homogeneous cartography of each parameter reported on one layer each one, along a large surface of 120 km long coastline. The computation is decomposed in different steps which consist from the qualitative factors, to quantify them and to normalize them in space. On the basis of field measurements and data analysis, four types of geological information have been added to the GIS model and a first computation of hazard modelling has been proposed to design a collapse hazard sensitivity map, based on a elementary summation of the parameters. We now prospect to introduce a ruled-based systems, dealing with the complexity of the interaction of the local parameters. An interaction network must be defined to represent the spatial and semantic links between local parameters.

Key words: GIS, rule-based engine, multiscale modelling, geology, risk analysis, general systems theory, feedback

1 Geological Context

Coastal chalk cliffs exposures along each part of the English Channel are composed of nearly vertical cliffs ranging from 20 to 100 m high, with a less or more thin cover of clays-with-flints and a chalk shore platform with a low angle of slope, often covered with sand and/or shingles. The shore platform is made of eroded chalk and is subjected to a semi-diurnal cycle of macrotides, whereas the cliff rocks are submitted to fresh groundwater that infiltrate within the chalk through rainfall (See Figure 1).

Coastal chalk cliffs exposed on each part of the English Channel suffer numerous collapses, with mean volumes varying between 10 000 and 100 000 cubic meters per event. Between October 1998 and October 2001, a minimum of 52 collapses have been observed along 120 km of the French chalk coastline located in Upper-Normandy and Picardy, with 28 collapses with volumes greater than 1000 cubic meters. Such collapses occurs inconsistently in time and space and appears to be relatively unpredictable. Little work has been devoted to the analysis of processes responsible for the collapses of the chalk seacliffs, and this led to the European scientific project, ROCC (Risk Of Cliff Collapse) because of the growing hazard to local communities from chalk cliff retreat. The main goal of the ROCC project was to identify the critical parameters leading to chalk coastal cliff collapses, and to evaluate the impact of those parameters and their interaction in such rock mass movements. The main objective was to create maps showing the sensibility of cliffs to erosion (cliff collapse hazard) along the 120 km coastline.

2 Cliff Collapse Process

The evolution of a cliff from stability toward failure, depends on changes present in the rock mass (lithology, fracture pattern), and processes acting within the rock mass (degree of water saturation, water movement) caused by external agencies of subaerial and marine origin. External agencies lead to :

- the development and opening of fractures (resulting from stress relief, fatigue, wetting and drying, freeze-thaw action),
- the deterioration of the rock material as a result of infiltration of water (resulting in solution, chemical alteration, physical breakdown through freeze-thaw or salt crystallisation),
- substantial geometric changes at the cliff foot (height of shingles and debris accumulation).

The rock mass characteristics such as chalk type, fracture pattern and karstic development may be consider as fixed parameters, only varying in space. Variable parameters such as water saturation of the chalk and water movement in the chalk through fractures and karstic system are closely linked to external agencies, with various delays. External agencies are varying in space and time, with various fitting scales. It is the case of climatological parameters, such as rainfall and temperature (with temporal variations from hours, days, seasons, years and decade) and oceanographic parameters with temporal variations from day to season (for tides and wave action), and variations from year to decade (for sea level variation due to global change).

3 GIS Approach

Cliff collapse process appears as a complex natural system, due to the large amount of parameters able to lead to a collapse. GIS approach has been used to allow an homogeneous cartography of each parameter reported on one layer each one, along a large surface of 120 km long. GIS has been also used to perform various combinations between each parameters to obtain various degrees of hazard.

The coastline location has been identified on the IGN topographic cartography basis at 1/25 000 (MNT IGN©1992, as a raster). Each GIS layer is dedicated to one parameter. The most simplified method to attribute a level to each parameter is to evaluate the minimal and maximal value of the parameter within a fixed geographic framework. The operator attribute the level zero for the minimal value and the value 100 for the maximal value. Then, each parameter (with various original units) may be reported on a layer with percents as a value of the parameter intensity ; moreover parameters may be combined easily together on the same geographic framework. The report of each parameter has been realised on a coastal strip located at the top of the cliff (drawn from a 50 m wide dilatation each part from the coastline). For each parameter, various strip sections (polygons) are defined as a function of the parameter value.

The GIS framework is usefull for the spatial correlations of various parameters and to combine several models to represent an hazard level.

4 Parameters Definitions

Georeferenced data produced by IGN have been used to build the GIS basis through Mapinfo® software. The various layers of the GIS are composed of parameters all varying in space, but non-variable or variable in time. Non-variable parameters in time are the geographic information (coastline location, cliff height) and the geologic information (chalk lithostratigraphic succession, fracturation). Variable parameters in time are the hydrogeologic and the oceanographic informations. Unfortunatly, oceanographic parameters have not been considered in this study. The cliff collapses occurrence results from the intercation of all these parameters, but we may consider that the observed location of past cliff collapses is a non variable parameter in time.

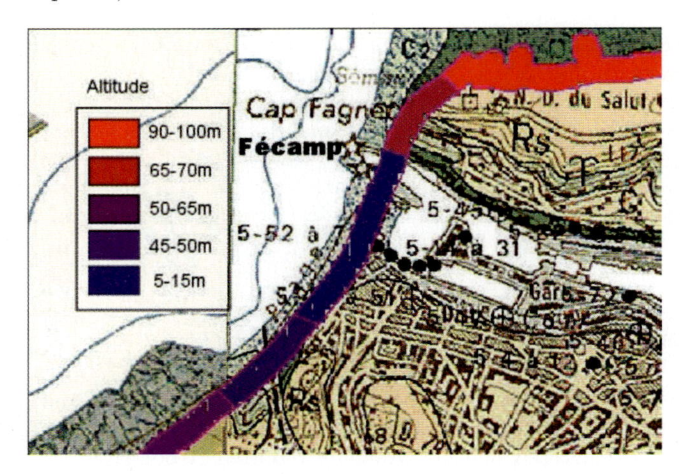

Fig. 1. Geological map (© BRGM) as raster around Fécamp and the strip indicative of the cliff altitude

4.1 Coastline location and cliff height

One layer is dedicated to the spatial location of the coastline (precisely defined at the top of the cliff) to build a 100 m wide strip.

A second layer is dedicated to the cliff altitude and has been used to select various height sections (as polygons) within the strip, with 10 m intervals, varying from 5 to 100m height. These layers have been deduced from the 1/25 000ème IGN topographic map (not shown in this study).

4.2 Past cliff collapses occurence

During the ROCC project, regular field surveys performed during october 1998 and december 2001 allowed to report a minimal value of cliff collapses. For each reported collapse, the location and date of occurrence were reported and the volume of the deposit were measured [Duperret et al. 2004]. Such data have been reported on a layer in the GIS, as recent collapses parameter.

4.3 Chalk lithostratigraphy

As defined in UK [Mortimore 1983], chalk type units are defined on the basis of a lithostratigraphic concept and are more representative of the geotechnical properties of the chalk than the stratigraphic scale traditionaly used in France [Mortimore 2001, Duperret et al. 2004]. The Chalk lithostratigraphy layer is composed of six chalk units of various characteristics detailed below :

1. Cenomanian craie de rouen is a nodular chalk with numerous flint bands in Upper-Normandy. Unit (1a) is defined between Antifer and Fécamp headlands, (1b) is defined north of Fécamp and (1c) south of Antifer.

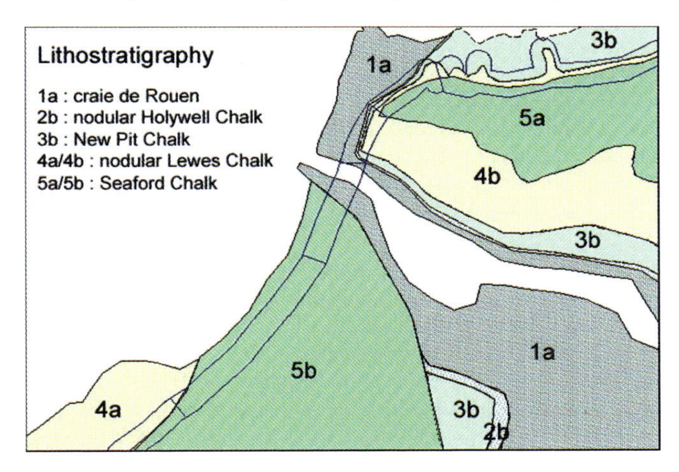

Fig. 2. Chalk Lithostratigraphy map, around Fécamp

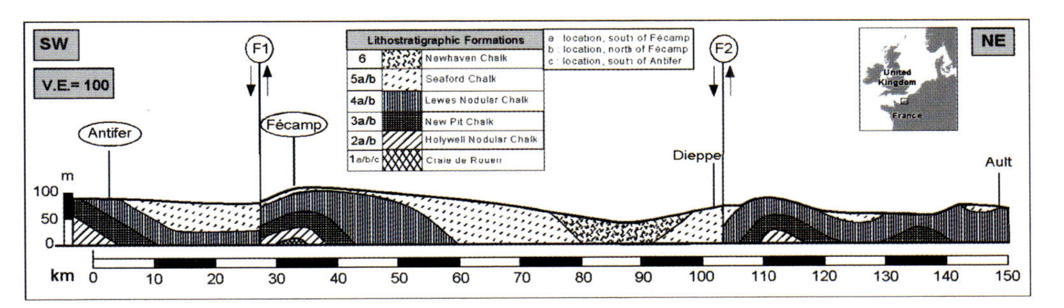

Fig. 3. Vertical cross section of the various chalk lithostratigraphic sequence on the cliff face

2. The Holywell nodular chalk is a nodular and massive chalk, with few flint bands, which contains many flaser marls and abundant Mytiloides shell debris layers, with open crossed fractures north of Fécamp (2a) and closed crossed fractures southward (2b)
3. The New Pit chalk formation contains numerous flint bands in cliffs south of Fécamp (3a) but is flintless northwards at St Martin-plage, north of Fécamp (3b),
4. The Lewes Nodular Chalk is a yellowish coarse chalk, including soft, marly bands and nodular hardgrounds, with regular flint layers. The Lewes Nodular Chalk formation contains dolomitic layers to the south of Fécamp (4a) which are absent northward (4b),
5. The Seaford Chalk Formation is a homogeneous white chalk with conspicuous bands of large flints, with large collapses, north of Fécamp (5a) and small collapses southward (5b)
6. The Newhaven Chalk Formation is a marly chalk with numerous marl seams and regular but few flint bands.

Chalk rocks present various degrees of physical properties, particularly density and porosity. The standart strength categories used to describe rocks for engineering purposes could not be applied readily to chalk, due to the variations in physical properties even within a single block of intact chalk. Even if each chalk unit is well defined, a direct comparison of each chalk units is not realistic. Moreover, some types of chalk retain water at saturation level while others gain and lose water more readily, changing drastically their physical properties.

The lithostratigraphic indice has thus been determined by a direct correlation between each vertical lithological succession on the cliff face and the estimated volumetric mass able to collapse. Each chalk unit formation overlay an older one. On the cliff face, various chalk successions may appears, depending on the height of the cliff and the thickness of each chalk units (Fig. 3). From field works, chalk units succession have been recognized on the cliff face outcrops, conducting to the definition of twenty four vertical successions of chalk units [Mortimore 2001,Duperret et al. 2004].

As examples : five chalk units outcrop on a vertical section at Fécamp (1a/2b/-3b/5b/5a) whereas one chalk unit outcrops at Dieppe (6). Cliff collapses volumes have been estimated for each event observed in the field in France and UK. The involved volumes are varying from 1 to 100 000 m^3 [Mortimore et al. 2004]. Seven volumic classes have been defined and each volumic class has been attributed to the corresponding lithostratigraphic sequence. On the basis of involved volumes during a collapse, a percent scale has been established for each lithostratigraphic sequence. The maximum influence has been established for the sequence 1a/2b/3b/4b/5a (at Fécamp) and sequence 4b/5a (100%), with collapses of mean volumes reaching 55 000m^3 and the minimum influence is established for the sequence 1b/2b (0%), with no reported collapse. The six other intermediates classes are based on a logarithmic scale and defined at 99%, 84%, 78%, 58%, 42% and 35%, with involved collapsed mean volumes of 50 000 m^3, 10 000 m^3, 5000 m^3, 500 m^3 and 100 m^3, 50 m^3 respectively.

4.4 Fracturation

On the basis of the observations in the field, a preliminary hypothesis was suggested that fractures embedded within the chalk cliffs could influence cliff collapse. About 2000 fracture orientation measurements were collected on 34 investigations sites regularly spaced along the 120 km long coastline. Fracture analysis were completed and homogenised on a systematic analysis of the cliff face from a continuous set of oblic aerial photography of the cliff face, realised in 1986. The correlation between field data and continuous aerial photography acquisition has been used to define two major layers dedicated to fracture occurrence [Genter et al. 2004].

Ransverse fracturation indice

The total number of fractures that appear on the cliff face represents in fact the number of transverse fractures to the cliff face. From this numbering, various sections of fracture density have been defined. The total number of fractures reported on a complete area of the cliff face vary from zero to 396 per square meter, which is equivalent to a scale of fracture density varying from zero to 0.17 and a mean space between fractures varying from zero to 95.5. An indice of transverse fractures has been calculated from these data, varying from zero to 100% and has been used to define 63 sections of various length on the strip, with various degrees of transverse fracturation.

Parallel fracturation indice

A second layer has been dedicated to fracture data parallel to the cliff face. Such data have been detected on aerial photographs on the naked beach platform, where fractures set appears easily. Calculations have been realised on a density fracturation scale with four levelsreported in percent in the GIS.

4.5 Hydrogeology

Experiments on chalk rocks show that hydration produces a marked decrease of the chalk strength, which varies depending on the chalk type. When chalk samples are submitted to a progressive water wetting, a fall of strength occurs. The decrease of the UCS strength is between 20 and 50% of the dry strength of chalk and this reduction begins with very low values of water content within the chalk [Duperret et al. 2005]. Chalk rocks formations are said to exhibit a dual porosity/permeability. In a classic dual-porosity aquifer the matrix pores provide storage and fissures provide the permeable pathways for flow. At large scale, the chalk aquifer presents a behaviour of a porous system, with low flow and at small scale, the chalk aquifer presents a behaviour of fissural system with high flows. At large scale, the water content of the chalk varies with fluctuations of groundwater level, submitted to rainfall inputs. The magnitude of the water table fluctuations in Upper Normandy is generally inversely proportional to the degree of fissuring (i.e. low permeability areas with less fractures have high water table fluctuations) [Crampon et al. 1993]. Hydrogeology data have been summarized on three layers in the GIS [Caudron et al. 2001].

Water table level

Data have been deduced from the hydrogeological map at 1/100 000 edited by BRGM. Original data was compiled from various available piezometric data in upper eptembe, that were acquired at various step of space and time. Even if this information is unprecise, it is able to give some interesting trends concerning the water table location in depth. Four classes of water table have been defined as a function of chalk imbibition thickness near the coastline : (0) zero for suspended valleys, (1) low imbibition thickness (lower than 5 m), (2) moderate imbibition thickness (around 10 m), (3) high imbibition thickness (higher than 10 m), (4) very high imbibition thickness (coastal area covered by impervious tertiary cover). The water table indice is ranging from 2.5 to 100 %, with the lowest water table effect for the class (1) and the highest water table effect for the class (4).

Coastal piezometric slope

Like water table layer, the coastal piezometric slope has been directly deduced from the hydrogeological map at 1/100 000. The degree of coastal piezometric slope gives an indication of the hydric flow, from the aquifer to the coastline. Five classes have been defined, with a piezometric slope ranging from 0, lower than 5 ‰, between 5 and 10 ‰, between 10 and 20 ‰, between 20 and 40 ‰, and higher than 40 ‰. The higher the slope is, the higher hydric flow is, the higher the influence to collapse is.

Coastal springs occurrence

Coastal springs locations reveal mainly the occurrence of fissural and/or karstic system in the chalk. These data have been collected in the field during eptember-october 1999 (fall 1999). Three classes have been defined depending of the springs flow and the spring density on a coastal section : (1) no coastal springs or coastal springs with low flow (lower than 10 l/s) and low linear density, (2) coastal springs with low flow (lower than 10 l/s) and high linear density or coastal springs with mean flow (between 10 and 100 l/s) and low linear density, (3) coastal springs with mean flow (between 10 and 100 l/s) or coastal springs with high flow (higher than 100 l/s). As a first approximation, the higher the flow and density are, the higher the karstic system is developed, and the higher the fissural flow transit is.

5 Hazard Modelling

5.1 Arithmetic combinaison

Based on the GIS model, a first computation of hazard modelling is proposed. The goal is to design a collapse hazard sensitivity map. The computation is decomposed in different steps which consist from the qualitative factors, to quantify them and to normalize them in space. The computation is based on a elementary summation of the parameters. The resulting sum in each coastal strip must be considered as a potential level of low to high degree of collapse hazard, based on a percent deduced from all parameters. A part of the resulting GIS is presented in the figure 4. Nevertheless, a confrontation with models and observations needs to be performed to introduce weighting in association with the pertinent parameters.

5.2 Multi-scale rule-based qualitative system

We now prospect to introduce a ruled-based systems, dealing with the complexity of the interaction of the local parameters. A process modeling allows to describe the alea estimation from complex interaction between geological structure, external agencies and hydrodynamic phenomena. The process modeling is described on Figure 5 where transitions are activate by physical qualitative modelling rules [Kuipers 1986]. During some specific external agencies (rainfall, temperature, ...), some links will be activated and dynamically propagate the phenomenon through the rules-based process which finally give in/output of the cliff collapse hazard. The resulting

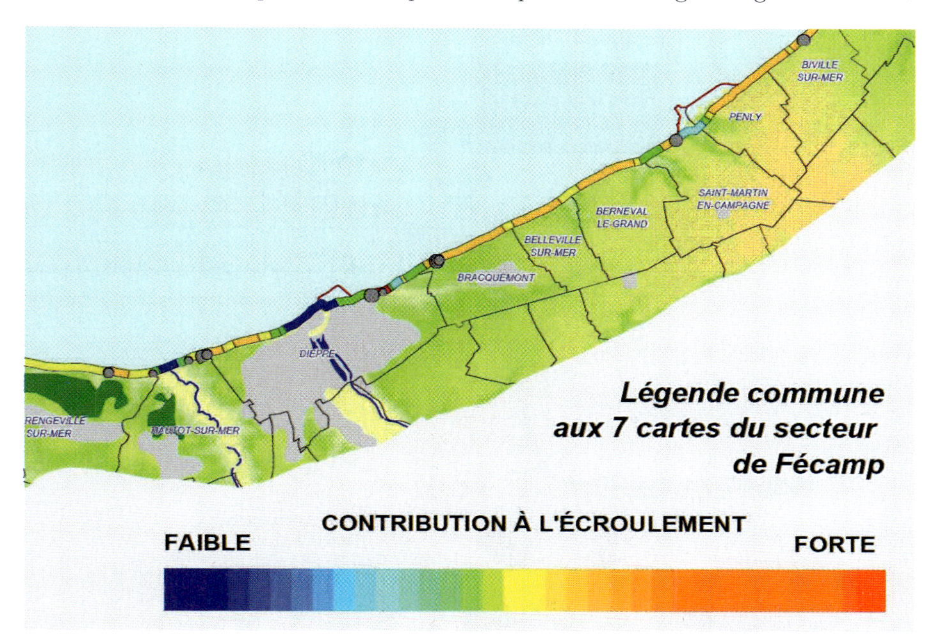

Fig. 4. Hazard Sensitivity map detail deduced from arithmetic combinaison between each parameters

collapse hazard can be considered as a kind of emergent computation. This process concerns a local description and the multi-scale approach consists to change the level of description and to represent each local process as a compartment. The compartments are linked by spatial transfers that correspond to pressure variation and water displacement along the fracture network as represented by the figure 5. The implementation of the whole method is in progress.

References

[Duperret et al. 2004] Duperret, A.; Genter, A.; Martinez, A.; and R.N. Mortimore (2004) *Coastal chalk cliff instability in NW France : role of lithology, fracture pattern and rainfall*, in : Mortimore R.N. and Duperret A. (eds) Coastal chalk cliff instability. Geological society, London, Enginneering Geology Special Publications, 20, 33-55.

[Duperret et al. 2005] Duperret, A.; Taibi, S.; Mortimore, R.N.; and M. Daigneault (2005) *Effect of groundwater and sea weathering cycles on the strength of chalk rock from unstable coastal cliffs of NW France.* Engineering Geology, 78, 321-343.

[Caudron et al. 2001] Caudron, M.; Equilbey, E.; and R.N. Mortimore (2001) *Projet ROCC : Hydrogéologie. Etat critique des connaissances et impact de l'eau sur la stabilité des falaises.*

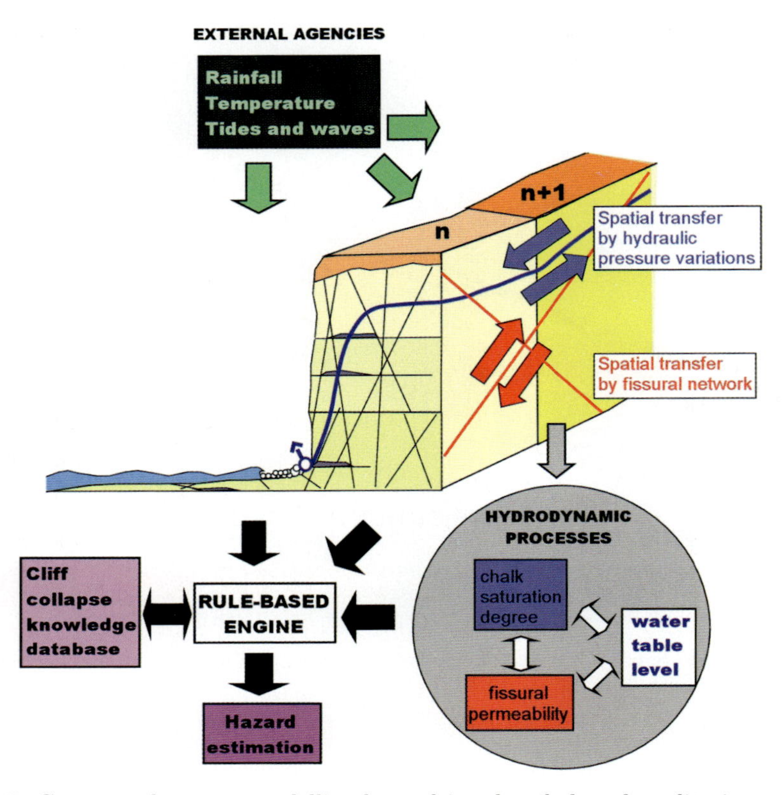

Fig. 5. Conceptual process modelling by multi-scale rule-based qualitative system

[Crampon et al. 1993] Crampon, N.; Roux, J.C.; Bracq, P.; Delay, P.; Lepiller, F.; Mary, M.; Rasplus, L.; and G. Alcayadé (1993) *France.* in : Downing R.A., Price M and Jones G.P. (eds). The Hydrogeology of the Chalk of North-West Europe. Oxford Science Publications, 113-152.

[Genter et al. 2004] Genter, A.; Duperret, A.; Martinez, A.; Mortimore, R.N.; and J.-L. Vila (2004) *Multiscale fracture analysis along the French chalk coastline for investigating erosion by cliff collapse*, in : Mortimore R.N. and Duperret A. (eds) Coastal chalk cliff instability. Geological society, London, Enginneering Geology Special Publications, 20, 57-74.

[Kuipers 1986] Kuipers, B. (1986) *Qualitative simulation.* Artificial Intelligence, 29, 289-338.

[Mortimore 1983] Mortimore, R.N. (1983) *The stratigraphy and sedimentation of the turonian-Campanian in the southern Province of England.* Zitteliana, 10, 27-41.

[Mortimore 2001] Mortimore, R.N. (2001) *Report on mapping of the chalk channel coast of France from Port du Havre-Antifer to Ault.* June-September 2001, Bureau de Recherches Géologiques et Minières (BRGM).

[Mortimore et al. 2004] Mortimore, R.N.; Lawrence, J.; Pope, D.; Duperret, A.; and A. Genter (2004) *Coastal cliff geohazards in weak rocks : the UK chalk cliffs of*

Sussex. in : Mortimore R.N. and Duperret A. (eds) Coastal chalk cliff instability. Geological society, London, Enginneering Geology Special Publications, 20, 3-31.

Structural and Dynamical Complexities of Risk and Catastrophe Systems: an Approach by System Dynamics Modelling

Damienne Provitolo

UMR 6049 ThéMA
Université de Franche-Comté
30 rue Mégevand
25030 Besançon cedex
damienne.provitolo@univ-fcomte.fr

Summary. Risk and catastrophe are complex systems. Within the scope of this paper, we focus our attention on structural and dynamic complexities of catastrophes and on the possibility of modelling and simulating its double complexity with a formal and methodological framework: the General Systems Theory and System Dynamics modelling. Then we briefly propose a model of urban catastrophe related to a flood. After we propose some ways of research allowing exceeding the limits related to the modelling.

Key words: Dynamic modelling, General Systems Theory, Risk analysis. Catastrophe, Complexity

1 Introduction

This document aims to apprehend the complexity of the systems of risks and catastrophes in urban environment I mean natural, technological, or even social ones. The catastrophe is a social and spatial disorganization of the territorial system which is affected by a disturbing event. Based on scientific research on risks and disasters and on our own research, we could identify various forms of complexity. Some of them concern the structural complexity of catastrophes, while others are related to the complexity of spatial and temporal scales. Others still depend on the complexity of geometrical forms and refer to the fractality of disasters. Finally, a last form of complexity is related to the non-linearity and the unpredictable dynamics of the systems. These various forms of complexity don't exclude each other, but can be observe together during a catastrophe. Within the scope of this paper, we focus first on structural and dynamic complexities of catastrophes and on the possibility of modelling and simulating its double complexity with a formal and methodological framework: the General Systems Theory and System Dynamics modelling. Secondly, we present briefly a model of urban catastrophe related to a flood. This model combines hazard, vulnerability and domino effects. It makes possible to apprehend the complexity of the relation man-nature and the nonlinear dynamics. Finally we propose some ways of research allowing exceeding the modelling limits.

2 Structural and Dynamic Complexities of Risk and Catastrophe Systems

Nowadays it is frequent to read in the literature that risks and catastrophes are complex phenomena. There are perceived and analysed from models which can be either politic, mediatic or scientific. The scientific models are rational models. In the past, the models of catastrophe were essentially focused on the study of hazard. Nowadays, the models try more and more to take into account the complexity of the disaster. Sciences of complexity propose a holistic approach to understand these phenomena. The holistic analysis seeks to understand systems mechanics by focusing not only on the entities which compose a system, but on the relations existing between these entities [Ménard et al. 2005]. Initially we will try to identify structural complexity of risk and catastrophe systems by undertaking this holistic approach. Two levels of structural complexities could then be identified. The first level depends on the even definition of catastrophe. This is defined by risk specialists as a combination, a conjunction of hazard and vulnerability ($R = A * V$). However we prefer to abandon this term of combination, which doesn't integrate interactions between constituents and define disaster as a complex set of hazard(s) (I mean the physical phenomena creating damages) and vulnerabilities (I mean the fragility of population, building, urban system). These two entities form the core constituents of catastrophe systems and their global functioning. In the absence of one or the other of these constituents, there cannot be a disaster. Thus, catastrophes are emergent phenomena of hazard and vulnerability. The emergence means that the global properties of the system cannot be deducted from the only knowledge of these entities [Zwirn 2005]. If we neglect an essential aspect of a complex system, we cannot understand the system in its whole. This means that the risk cannot be reduced neither to the one or other one of its parts nor to the sum of its parts. This is an important finding. For a long time, the terms of risk and hazard were used as synonyms, especially in the field of natural risks, such as floods and earthquakes. As a consequence, scientific research widely privileged the study of hazard. Of course, scientific progress in this domain was particularly important. In fact, by neglecting the vulnerability entity and therefore omitting a part of the system, important aspects of the structure and the global behaviour of catastrophes were missed out. The sciences of the complexity thus developed the concepts of risk and catastrophe.

The other type of complexity is relative to the increase of the levels of complexity when we go from some sector-related complex risk (I mean hazard and vulnerability; for example, a seismic, flood or technological risk) to transversal complex risk (again I mean hazard, vulnerability and "domino effects") [Dauphiné 2003a]. Domino effects are a chain of events activated by hazard or vulnerability. For example, a seismic risk releasing a technological risk or even a flood risk releasing acts of panic. This increase of the levels of complexity results not only from the even higher number of constituents but also and especially from the multiplicity of the interactions which unite the various entities of a catastrophe. So, the transversal complex risk integrates the varied nature of "domino effects": natural and technical or technological, natural and social or still, to take a last example, natural, technical and social. These "domino effects", particularly in urban areas, are creative of clearing of multiple thresholds of gravity in varied domains and for the same "hyperdisaster" [Guihou et al. 2006]. This structural complexity is rarely taken into account by the managers of disasters, which often study only one category of disaster: the risk of floods, earthquakes, forest fires, tsunamis, nuclear accidents etc. are identified in the various documents of prevention and management of the major risks (Municipal Document of Information on Major Risks). But these various documents don't integrate the transverse complexity of catastrophes which take place in urban areas. Some exceptions exist in this domain, notably in Japan, where the authorities of Tokyo are afraid of a chain of natural and technological disasters following an earthquake [Hadfield 1992].

The Cartesian approach, which is limited mostly to the analysis of one type of risk, does not support the understanding of all the mechanics of a disaster. This scientific method tends at present to be replaced by holistic approaches. For the better understanding of a catastrophic event, the study of relations and interactions is better than the study of the system constituents.

The second form of complexity depends on the dynamic and unpredictable complexities of the disaster systems. The General System Theory of Ludwig von Bertalanffy [Von Bertalanffy 1973] and the systemic modelling offer a formal frame to build models of disasters which are centred on the complexity of the relations man / nature and on nonlinear dynamics. Within the framework of the General System Theory, the System Dynamic of J.W. Forrester [Forrester 1984, Aracil 1984] is interested in the changes which occur inside the studied systems. This method of modelling and simulation of complex systems appeared in the early 1960s when J.W. Forrester was a professor at "Sloan School of management" of MIT (Massachusetts Institute of Technology). They were developed in France from 1980, with the translation into French of his book "Principles of Systems". The models allowing apprehending the dynamics of a system base themselves on the concepts of interaction, feedback loops and complexity. The dynamic and unpredictable complexities of catastrophe systems result from feedback loops (a circle of cause-and-effect) which connect the different variables of the system, the interactions between these feedback loops and the delays between variables and nonlinear phenomena [Donnadieu et al. 2002]. Two kinds of feedback relationships govern the dynamic systems of a catastrophe. Some push the system towards instability and disorder. These are positive feedback loops. Others make the system return to its initial state. These are negative feedback loops. These two kinds of feedback relationships operate in complex systems. The presence of these circuits of feedback loops, which can occur either simultaneously or successively, does not allow, without modelling and simulation, to predict the temporal dynamics of a system. This dynamic complexity will be increased by the presence of a temporal gap and non-linearity between variables (no proportionality between cause and effect). In the event of a disaster, the planning and the intervention of disaster managers are forces which, normally but not systematically, bring the system to stability and order. On the other hand, negative forces can push the system to disorder and to instability.

It is necessary to apprehend the catastrophes as a system with several constituents and which evolution is rarely linear. The System Dynamics modelling was thus retained to build models of urban disasters. The construction of a model starts by the realization of a Forrester diagram. This diagram represents the various elements which compose the system in term of stocks and flows, and clarify the relations established between the various variables of the modelled system. It is what we call the "Dynamo language". The relations between the variables of the system are mathematically formulated with for example statistical laws, logical rules (if then else). In this way quantitative and qualitative data can be integrated into the model. The basic elements of the model are the state variables, flows, converters and connectors. It is difficult to separate this approach of the computer tool allowing its application.

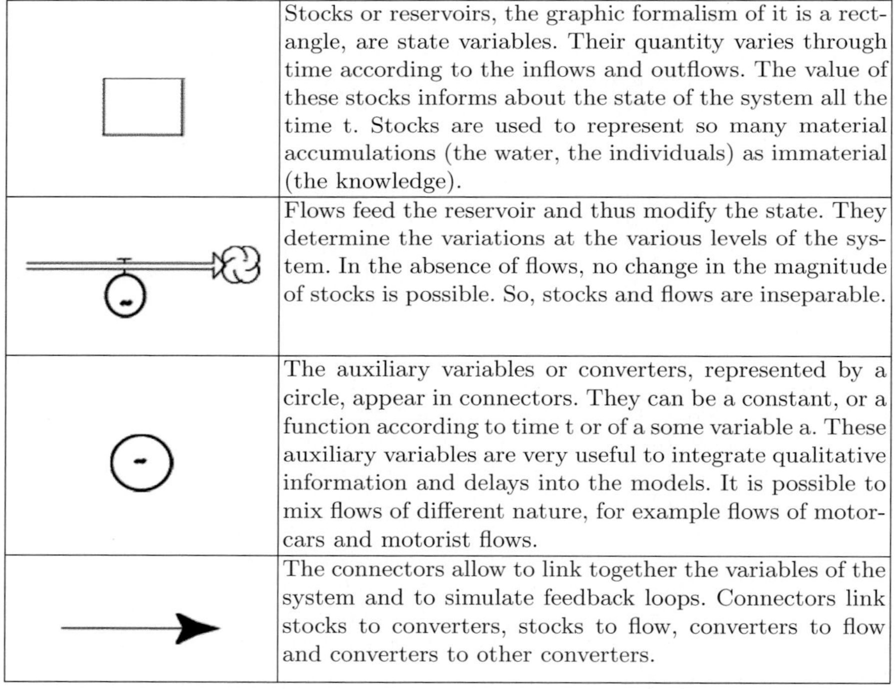

	Stocks or reservoirs, the graphic formalism of it is a rectangle, are state variables. Their quantity varies through time according to the inflows and outflows. The value of these stocks informs about the state of the system all the time t. Stocks are used to represent so many material accumulations (the water, the individuals) as immaterial (the knowledge).
	Flows feed the reservoir and thus modify the state. They determine the variations at the various levels of the system. In the absence of flows, no change in the magnitude of stocks is possible. So, stocks and flows are inseparable.
	The auxiliary variables or converters, represented by a circle, appear in connectors. They can be a constant, or a function according to time t or of a some variable a. These auxiliary variables are very useful to integrate qualitative information and delays into the models. It is possible to mix flows of different nature, for example flows of motorcars and motorist flows.
	The connectors allow to link together the variables of the system and to simulate feedback loops. Connectors link stocks to converters, stocks to flow, converters to flow and converters to other converters.

Table 1. Presentation of the main symbols of the graphic module of the software Stella Research

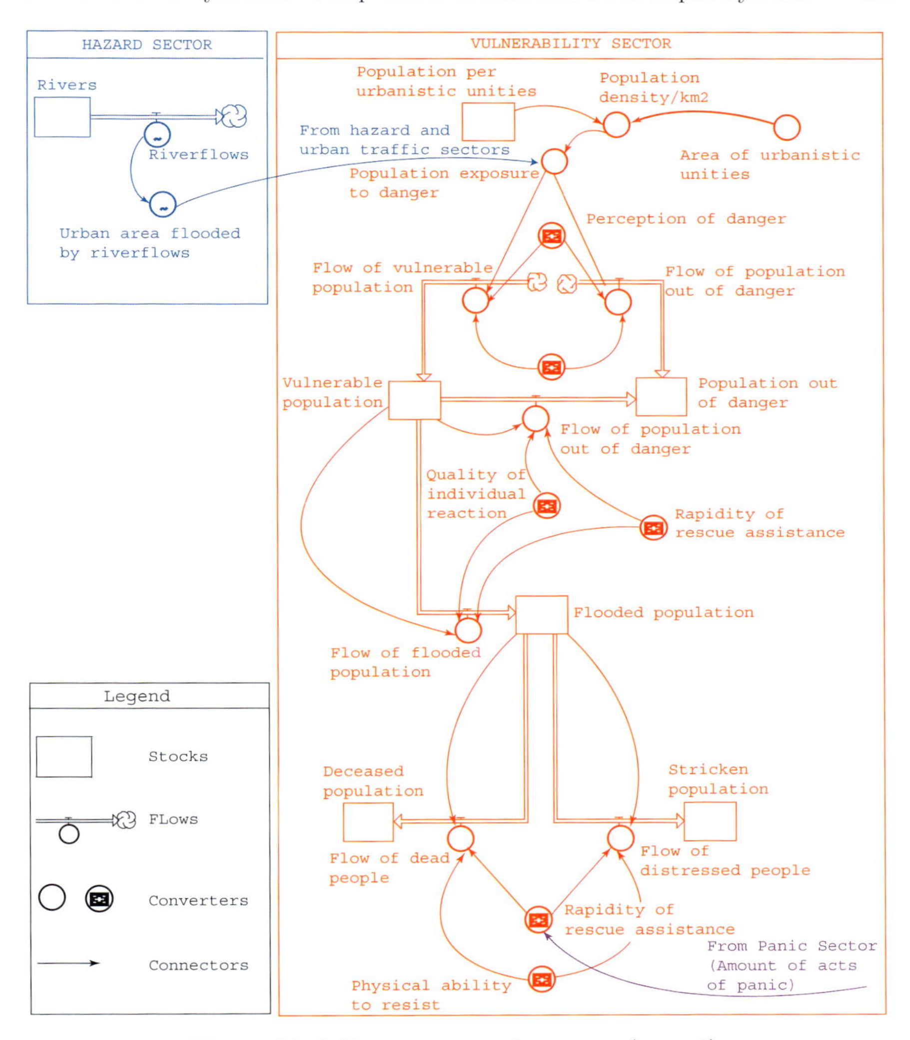

Fig. 1. Modelling a catastrophe system (part. 1)

3 System Dynamics Modelling of Catastrophe with the Stella Research Program

At the moment, Stella Research [Stella Research 1997] is one of the most wide-spread software to model and simulate the complex systems according to the principles of a Forrester formalization. This software includes two modules: a graphic module which supports building the structure of the model. A mathematical module presents the results in the form of curves which is a set of differential equations defined by the graphic module. These differential equations are discretized in difference equations

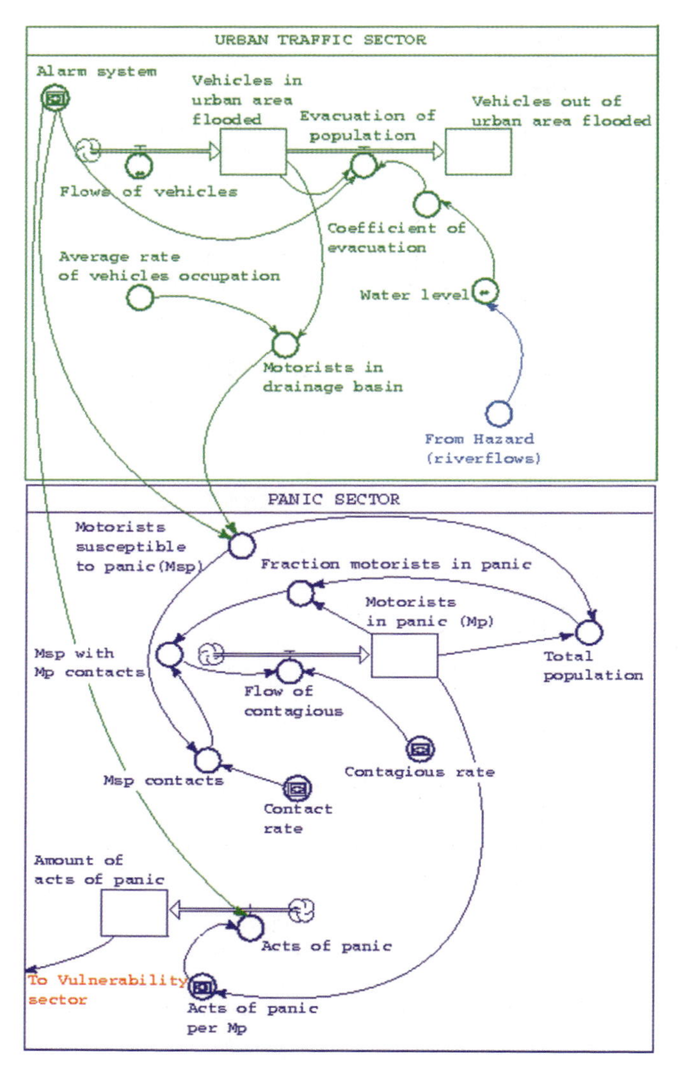

Fig. 2. Modelling a catastrophe system (part. 2)

in the software Stella Research. The main symbols of the graphic module of the software Stella Research are represented in the table 1.

This software was used to create a model of urban disaster of natural origin, a flood. This model (figures 1 and 2) has already been the object of publications [Provitolo 2007, Provitolo 2005]. We invite the reader interested in greater details of the structure and the results of the simulation to refer to these publications. The originality of this model is to associate various entities of disaster and its "domino effects",

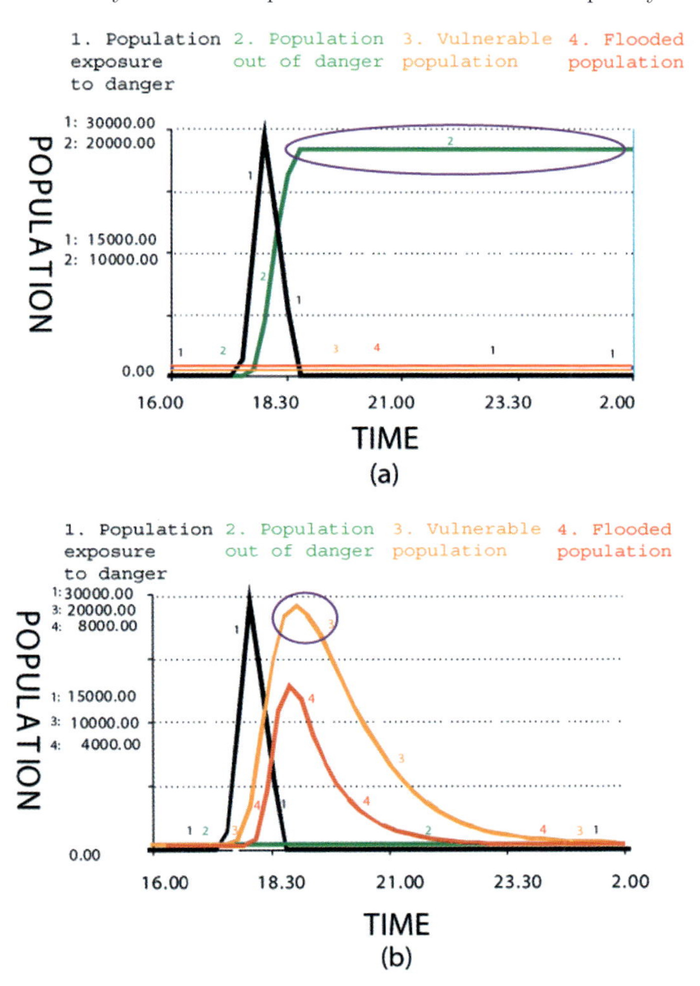

Fig. 3. An example of simulation of the population vulnerability during a catastrophe

namely the flood hazard (in this case, we have only considered a river flood), the vulnerability of the population during the flood, the problems of urban traffic and of motorists evacuation as well as a sector of panic. We find here two forms of complexity. We have a first form of complexity depending on the interaction between the various elements of the system (I mean interactions intra and inter-sectors). And we have a second form of complexity depending on nonlinearity. The system of catastrophe is in this case constituted by four sectors in interaction. The sub-system of hazard establishes a link with the sector of the population vulnerability subject to a flood. Indeed, the exposed population is a function of the area (km2) flooded by river flood (sector hazard). This hazard, physical phenomenon at the origin of the damages, is also going to have impacts on the urban traffic. It can create panic behaviour amongst the motorists taken by the streams. These acts of panic, which can lead to acts of abandonment of vehicles, go then retroact on the sector of the vulnerability by delaying the arrival of emergency services. This global model of disaster thus associates aspects generally treated independently of each other. Every sub-system is constituted, under graphic shape, by state variables, flows, converters and connectors. These graphic models are then transformed into computer models which allow running simulations.

The results of simulations are curves which show the temporal dynamics of the catastrophe. According to the values of the various parameters of the model (policies of prevention, rapidity of rescue assistance, rate of contact and contagious of the panic), the forms of the dynamics of the catastrophe system modify. So, the system knows different evolutions. It does not adopt the same behaviour. Also the simulations lead to results which are counter-intuitive, unpredictable: for example, the decrease of the rate of contact and contagious building the relation between motorists susceptible to panic and panicking motorists does not decrease the stocks of "contaminated" population (in this case motorists in panic). With nonlinear dynamics, the system can thus adopt unpredictable behaviour [Provitolo 2005]. In the same way, we can observe that, the variability of the factors of vulnerability create qualitative change in the vulnerability dynamic (figure 3). For a good prevention (a), we go from constant curves and curves equal to zero to some curves with positive dissymmetry for a bad prevention (b). We observe threshold effects and rupture effects.

These results of simulation also have the benefit to reveal that the complex dynamics of a system are not only read in the result of complex curves, for example chaotic curves. The complexity results of non linear equation and interactions between variables. But this equations and interactions do not create necessarily complex curves. In fact, few results of simulation reveal complex curves. These results of the simulation, by giving various possible evolutions of the system, are a tool of understanding. They also impose a rule of conduct. The models have to be the object of simulations based not only on average but also on extreme values of parameters. It is the most effective way of observing the variety of trajectories.

4 Conclusion

This System Dynamics modelling offers a formal and methodological frame to apprehend the complex structures of types of catastrophes (natural, technological or technical and social) and different kinds (flood, earthquake, terrorism) as well as their temporal dynamics. This modelling allows analysing the catastrophes in its whole. However this approach has limitations, notably with respect to the consideration of the space. The space is indeed indirectly integrated into the models, for example by means of densities or of surfaces. To exceed this limitation, the "systemicien" can use other tools of modelling. The scientists have two solutions to integrate space in all its heterogeneousness and realize models which take into account temporal and spatial dynamics of a catastrophic event. The first solution is to link a Geographic Information System (GIS) with a model of system dynamics to integrate the spatial constituent. Currently, the GIS allows combinatorial analyses and structures, but on the other hand is less capable of modelling and simulating dynamics and complex dynamics. It is very likely that these limitations will be overcome within the next years. The second approach is to combine System Dynamics modelling with cellular automata. This research is in progress at the University of Maryland. The philosophy of this spatio-temporal modelling is to integrate the spatial systems, the simulation and the complexity into the problem of risk. So we could have the possibility of working on a spatial grid of square unit cells (raster) representing an urban area. In each of these cells a model of System Dynamics of catastrophe (stock-flow model) would be connected. The same structure of model would run in each cell. In this way we would obtain interactions between the cell and the systemic model, namely between the urban shape and the variables of the catastrophe system (hazard, vulnerability and "domino effects"). This architecture of models in interaction would allow understanding not only the spatial dynamics but also the temporal dynamics of the catastrophe. Those research efforts are certainly worth being further developed. Indeed, the knowledge obtained by these simulations would certainly allow taking up new forms of management (how to simply manage complex phenomena), and would make possible the realization of a comparative analysis of risks and catastrophes.

References

[Aracil 1984] Aracil, J. (1984) *Introduction à la dynamique des systèmes.* Presses Universitaires de Lyon, Lyon, 412 p.

[Von Bertalanffy 1973] Von Bertalanffy, L. (1973) *Théorie générale des systèmes.* Dunod, Paris, 296 p.

[Dauphiné 2003a] Dauphiné, A. (2003) *Risques et catastrophes.* Armand Colin, 2ème édition, Paris, 288 p.

[Dauphiné 2003b] Dauphiné, A. (2003) *Les théories de la complexité chez les géographes.* Economica, Paris, 248 p.

[Donnadieu et al. 2002] Donnadieu, G.; and M. Karsky (2002) *La systémique, penser et agir dans la complexité.* Ed. Liaisons, Rueil-Malmaison, 269 p.

[Forrester 1984] Forrester, J.W. (1984) *Principes des systèmes.* Presses Universitaires de Lyon, Lyon.

[Guihou et al. 2006] Guihou, X.; Lagadec, P.; and E. Lagadec (2006) *Les crises hors cadres et les grands réseaux vitaux- Katrina.* Mission de retour d'expérience, EDF, 34 p.
`http://www.patricklagadec.net/fr/pdf/EDF_Katrina_Rex_Faits_marquants.pdf`

[Hadfield 1992] Hadfield, P. (1992) *Tokyo séisme : 60 secondes qui vont changer le monde.* Ed. Autrement, Paris, 149 p.

[Ménard et al. 2005] Ménard, A.; Filotas, E.; and J. Marceau (2005) *Automates cellulaires et complexité : perspectives géographiques.*
`http://www.iag.asso.fr/aarticles/AUTOMATES`

[Provitolo 2007] Provitolo, D. (2007) *Vulnérabilité aux inondations méditerranéennes en milieu urbain : une nouvelle démarche géographique.* Annales de Géographie, 653.

[Provitolo 2005] Provitolo, D. (2005) *Un exemple d'effets de dominos : la panique dans les catastrophes urbaines.* Cybergéo, 328.

[Stella Research 1997] Stella Research (1997) *An Introduction to Systems Thinking.* Hanover: High Performance Systems.

[Zwirn 2005] Zwirn, H. (2005) *Qu'est ce que l'émergence.* Hors-Série Sciences et Avenir, juillet-août 2005.

Detection and Reification of Emerging Dynamical Ecosystems from Interaction Networks

Guillaume Prévost and Cyrille Bertelle

LITIS, University of Le Havre
25 rue Philippe Lebon, BP 540
76058 Le Havre cedex, France
cyrille.bertelle@univ-lehavre.fr

Summary. In this paper, we present an hybrid ecosystem modeling based on emerging computation from interaction networks. Initially based on an individual-based modeling (IBM) simulation, we propose an automatic computation to detect predator-preys systems. After their detection, these systems are replaced by a differential system during the simulation. In this way, we can change the description level and improve both the computation time and the whole system analysis by detecting some emergent organizations. The description modification between IBM representation to differential one needs to identify the global coefficients of these differential equations. Due to the complexity of relations between these two kinds of representations, a genetic algorithm is proposed to solve this identification.

Key words: Complex systems, ecosystems, interaction networks, emerging systems, non-linear differential systems, genetic algorithms, ontology.

1 Complexity as a Matter of Interactions

1.1 Emergence of complex systems

In 1968, L. Von Bertalanffy [Von Bertalanffy 1968] describes systems as a set of elements in interaction. Thus, the becoming of an element is strongly linked with its local environment. Eventually, some elements in interaction become related in dynamics and therefore constitute a global entity at a higher level. Finally, entities evolve at their own level and modify their parts.

Recently, most scientists describe the complexity as a result of interactions between elements of a system. They point out the fact that those systems are non-deterministic ones [Cardon 2004]. This fact implies that complex systems cannot be studied by dividing them into smaller parts but as a self-organizing systems [Cardon 2005] in which smaller systems sometimes emerge. Those systems stand for local sets of elements temporarily interrelated in dynamics which can aggregate or reject elements (or totally disappear) as they undergo new interactions from their local environment. The General System Theory [Le Moigne 1994] sketches the basis of the complex systems structure.

As a consequence, we can describe complex systems as a system according to L. Von Bertalanffy. The structure of that system evolves (depending of the local interactions between its parts) on different scales of time and space. This structure is made of entities (or emerging systems) appearing and disappearing depending on the configuration of the interactions network of its elements. To conclude, the structure of a complex system results of the feedback between its parts on their own scale and between the scales. A complex system is made of complex systems, the whole influencing the parts and the parts modifying the whole.

1.2 Modelling: an interaction network based approach

As we previously described, complex systems embody the rise and the disappearing of systems. Since many years, models are being developed to study the evolution of those systems. In ecology and especially in population dynamics, a great diversity of tools coming from mathematics, computer science or medicine for example are used to describe the interrelated becoming of populations in their environment. Each model can be analyzed by considering the associated interaction network between its elements [Prévost 2005].

Let us give a simple example with a very classical model: the logistic growth of a population. This model only embodies the dynamics of one population limited by an external factor which can be a critical resource. Therefore, the interaction network is simply made of the population and its resource linked by the consumption interaction. When the resource is too low, the consumption is decreasing and the growth is weakened. More complicated models induce more complicated interaction networks but it is important to underline the fact that, in particular conditions, one can associate a model to a network.

Some models directly express their associated interaction networks by their forms. For example, linear or non linear differential models (using one or more equations) mostly express the interactions as a term of one of their equations. Matrix models (as Leslie ones) fix the interactions between the populations in their values.

1.3 Using Individual Based Models to study interaction networks

Since its definition in [DeAngelis et al. 1992], Individual Based Models (IBM) have been widely used in many sciences. Questions remain regarding the use of the results [Grim 1999]. Basically, IBM differ from global models as they focus on individuals and not directly on populations. Thus, both of them is relevant at a different scale and apply to different kind of studies. IBM are not only a discrete expression of global models but could induce non-deterministic aspects as individuals are in a local context. We can assimilate individuals in IBM as elements in our ecosystem[1].

The interactions modeled in IBM are on a lower scale than interactions in classical models. Despite this fact, if one considers the interaction network between the individuals, one can underline the fact that this interaction network corresponds to a more global one. In complex systems, systems keep on emerging and disappearing. Those systems correspond to temporary global entities and therefore can be modeled thanks to a global model. As we will introduce, we can detect in IBM temporary stable interactions networks corresponding to global models.

In the following, we will first describe the complexity of ecosystems to propose an holarchic hybrid model. To deal with the complexity description of knowledge involved and developped in our model, we use the concept of ontology in order to define a general ecosystem model and we will present a distributed implementation of the resulting IBM simulation. We will then focus on the illustration of complex properties by highlighting the need of specific interaction between individual and environment. We develop then, a genetic algorithm allowing to change the level description of the systems, from IBM emergent behavioral systems to Lotka-Volterra equational systems.

2 Ecosystems Are Complex

The botanist Tansley [Tansley 1935] defines ecosystems as "The more fundamental conception is ... the whole system (in the sense of physics) including not only the organism-complex but also the whole complex of physical factors forming what we call the environment. We cannot separate them (the organisms) from their special environment with which they form one physical system ... It is the system so formed which [provides] the basic units of nature on the face of the Earth ... These ecosystems, as we may call them, are of the most various kinds and sizes."

Typically, ecosystems are described as a biotope and a biocenisis in mutual interaction. Moreover, they are crossed by fluxes (mass transport or energy) which structure them in dynamical way. A reductionist approach fails in modelling such mutual interactions and feed-back processes. New approaches based on general system theory concepts have therefore been tried to produce more efficient models.

Ecosystems are systems as described in the general system theory [Le Moigne 1994] and can be seen as a set of interacting elements which are characterized by the following aspects:

- **Mutual dependence.** Each element is directly linked with other elements in structure or dynamics. Therefore, its evolution depends on the other elements in interaction with it. Finally, separating an element from its neighborhood modifies the whole system.
- **Emergence of organizations.** The interaction of elements leads to the emergence of natural organizations which generate "new entities". Those entities differ from their components in their structure and dynamics.

[1] Of course, they are not the only elements.

- **Feed-back processes.** This is the retroaction from the emergent organizations to its own components.

Emergence processes act recursively and generate hierarchical systems organization. An adapted description is the concept called SOHOS, according to Koestler [Koestler et al. 1969]. SOHOS stands for Self-Organized Holarchic Open Systems. A holarchy is a non-directional hierarchy in which the members are called holons.

If we consider the three previous characteristic aspects of ecosystems, a suitable definition of ecosystems could be: "biotope+biocenosis", natural multi-level holarchic systems, crossed by structuring fluxes.

Ecosystems as holarchic systems, must be studied at many levels of time and space (each level is significant) and are crossed by fluxes. Many tools exist to model and simulate ecosystems. These tools differ by the nature of the model they use, the level they describe and the phenomena they directly take into account. Thus, they mainly focus on one aspect of the ecosystem. Therefore, gathering all these models is a hard task due to structural differences between these tools. Mixing these models is seldom theoretically feasible, and when it is, the computation of the resulting simulation needs an impressive computation time. Unfortunately, raising the number of modeled interactions is the concern of most modelers. We aim to provide a method facilitating the mixing of models in order to simulate aquatic ecosystems in a multi-scale description.

3 An Holarchic Hybrid Model for Ecosystems

An ecosystem life cycle always goes through three temporal states characterized by its biotope and its inner complexity.

- At the beginning, the profile of an ecosystem can be described as juvenile. The ecosystem itself contains many raw materials and its biodiversity is low. The biotope is mainly made of simple organisms. These organisms modify the layout as they keep on multiplying and consuming the raw material. The environmental condition has low influence on their development. The main factor limiting their growth, in size and in number, is the quantity of material and the space.
- Then, an ecosystem evolves into a mature or adult form. An adult ecosystem undergoes a replacement of its original settlement (simple organisms) with complex organisms. These organisms suit the ecosystem characteristics and tend to maintain them. They consume and produce (or participate to the production of) raw materials. Those organisms replace some simple organisms (but some simple organisms remain). So on, the biodiversity of the ecosystem is very high. We should be aware that the complex organisms need the building made by the simple organisms during the juvenile stage to appear.
- Finally, an ecosystem tends to be aging. When being adult, the ecosystem is made of many types of complex systems, each one participating to the maintenance of the system. As it grows old, the ecosystem tends to lower its complexity and biodiversity by the elimination of the less performing complex organisms and by the conservation of the best organism association.

During its life, an ecosystem always undergoes stress periods making it pass from one step to another (sometimes regressing as it is the case when an ecosystem is being exploited). We point out that biotope and biocenosis are directly linked and that each organism acts for his ecosystem as the same time as it is influenced by it.

Following this line of thought, we adopt a well-known description of the biotope based on a classification of the organisms. This classification is made of three parts:

1. **Producers.** Producers are the base of the ecosystem. They are responsible for the production of organic mass by consuming inorganic materials (mineral salt for example). Moreover, they produce many substances and heat. They too are eaten by consumers.
2. **Consumers.** An ecosystem most of the time is studied through its organic mass. In an ecosystem, most organisms try to develop and support its organic mass. This category of organism is consumers. They mostly produce organic mass by consuming other organic mass and release part of that mass in the ecosystem. They too participate to the layout of the environment.
3. **Detritivors or decomposers.** As an organism lives, it releases (dead) organic mass. This organic mass is reused by bacterias to produce raw materials needed by the producers. They thus maintain the ecosystem resources.

Each part of this classification is a part of the functional aspect of the ecosystem and participate to the previous cycle of life. This classification constitutes the base of the representation of the biotope in our model.

We represent ecosystems with a multi-level model. Some level are fixed “a priori”. We shall describe them later. The determination of the level lays on the following concepts.

Ecosystems are systems as described in the general system theory meaning they are made of elements in interaction. In ecology, the basic elements are individuals. So on, we should introduce individuals in our model. Clearly, we should use an Individual Based Model on that level. Moreover, ecosystems are thermodynamical systems. We should take into account flows between ecosystems on a different level than the individual's one (as they operate on a different scale). Finally, those flows are well modeled by laws. Finally, ecosystems are SOHOS. That induces we shall define non-directional relations between levels and furthermore, and add dynamic scaling for emerging entities.

While studying the informations available concerning several natural aquatic ecosystems (the Seine estuary, for example), we noticed that these ecosystems could be separated into different compartments, each one being a particular ecosystem. Clearly, that description is space-oriented depending of the localization of the compartment and its inner space. A level corresponding to those compartment should be adopted.

Definition of the holarchic hybrid model

Following the previous concepts, we propose a model suiting ecosystems features and allowing reuse of already existing model different by nature (law or rule-based, different scales).

The model is individual-based in the way it represents the different entities of an ecosystem. As a consequence, each entity has states (e-states, i-states or p-states) and behaviors. The behaviors correspond to the model used to model the dynamics of the entity. It could be rule-based or law-based.

Our model presents three levels defined "a priori".

- **Individual level.** This level is the lower one in the holarchy of our system. It embodies entities that could not be decomposed. We introduce individuals and super-individuals at this level. Moreover, the elements of that level are described following the consumer-producer-detritivor model. Thus, one should question what are the links between a particular entity and the ecosystem. That level is clearly individual-based so behaviors of the entities are rule-based and entities have i-states.
- **Compartment level.** The space is a critical data in ecosystem modeling. We introduce the space at this level. Thus, a compartment is a single entity in interaction with other compartment (exchanging flows) and containing individuals. As ecosystems are SOHOS, individuals and compartments influence each other. The compartments have law-based behaviors and e-states. The e-states of the compartment correspond to global values considered homogeneous in all the space of the compartment.
- **System level.** Systems are non-spacialized entities corresponding to a set of entities. The link between these entities could be defined "a priori" or during the simulation. The first example of such a system is the ecosystem itself. It is made of compartments and defines their relations. Moreover, phenomena occur at the scale of the ecosystem itself influencing all the compartments. The systems have e-states and law-based behaviors. During the simulation, different kind of systems could emerge, each one corresponding to a new scale. Moreover, phenomena with particular scales should be modeled through systems. It is important to understand that the phenomena occurring on the ecosystems level directly modify the compartment and thus influence too the individuals of the compartment themselves. Simultaneously, individuals modify their compartment and thus influence the ecosystem.

What will be modeled?

The model developped in the following is an ecosystem in which we study the influence of light and oxygen on a simple food chain. The light differs in many places of the ecosystem. The oxygen influences the behavior of the biotope.

The food chain is made of four species. The first population is made of planktons consuming mineral salt and producing oxygen depending on the light. Bacterias constitute the second one. They decompose organic mass and release mineral salt. The two last populations are fishes. The first of them feeds on planktons. The second have plankton and the first population of fishes as preys. Both of them consume oxygen.

4 Ontology of the Ecosystem Modelling

4.1 Ontology as a medium of engineering.

The model itself, despite the fact that it contains implicit informations, cannot be considered as an efficient way to communicate the knowledge about how to make a model corresponding to a case of study. In order to do so, we make an ontology of that model which constitutes a meta-model.

Ontology [Gruber 1991] stands for clear definition of a domain of knowledge respecting the following criteria :

1. Ontology must be clear and splits the domain of knowledge in many concepts that are related. Concepts are linked and the label of the link describes the way they are in relation. A concept can be a particular case of another. In this case, the latter inherits from the former. Every concept and link must have a clear definition.
2. Ontology must be sharable. In fact, one should be able to use concepts or links for its particular domain of knowledge. For example, basic concepts of ecology are common to different domains of knowledge or one should want to apply a topology of links to its particular domain of knowledge which has nothing to do with ecology.
3. Ontology can be completed meaning users should be able to add new concepts to the ontology for it to fit their needs.

All those features fit with the needs of model engineering so we made an ontology of our model using Protege. That ontology explicits the structure, interactions and feedbacks described in the model as well as its inner way to relate models and levels. Moreover, it is linking the model with concrete elements from ecology helping users to apply the model to their case of study. For example, we introduce a decomposition of populations between producers, decomposers and consumers and link it with the different models of our meta-model. Therefore, we can complete the meta-model's range by increasing the concepts of this part of the ontology.

Finally, ontology are not a static view of a domain of knowledge but should be applied to particular usecase. Doing so the knowledge of the particular case applies and therefore helps us understand it and see how all the parts of this usecase are interrelated. That process is sometimes called instantiation as it consists in giving concrete values from our usecase to all the linked concepts of the ontology.

4.2 Presentation of the ontology in Protégé

Particularity of Protégé

Protégé is a well-known freeware commonly used in computer science to conceive ontology and exploit them. It adopts an object way to represent concepts and links. Therefore, every concepts stands for a class with a clear definition of the concept it is representing. In addition to that definition, particular attributes have to be added by the user. Those attributes are called slots and constitute the links between concepts. Thus, when a concept A is linked with another concept B, a slot is defined in the slot A which type is the class representing concept B. The name of the slot describes the link itself and a clear definition must be added. Additional features of slots exist as the cardinality or the fact that the link is an inverse one.

Finally, making an ontology in Protégé consists in making a class hierarchy where classes are linked.

The ontology itself

The ontology is divided in four parts representing particular concerns about modeling and whose elements are linked.

Entity stands for every type of data represented in the model. It is named and linked with mathematical values that are the states of the data (for example, height or number). Therefore, particular subclasses of entities have been defined, describing particular ways to model data at a particular scale:

1. Global entities represent non localized elements described by continuous values on a global scale (for example, settlements).
2. Individual gathers all ways to model discrete entities in a local space (see figure 1). Therefore, it is divided between single individuals which represent every individual as a particular entity and super-individuals, each of which regroups some individuals that are localized at the same place and have equal states [Scheffer et al. 1995].

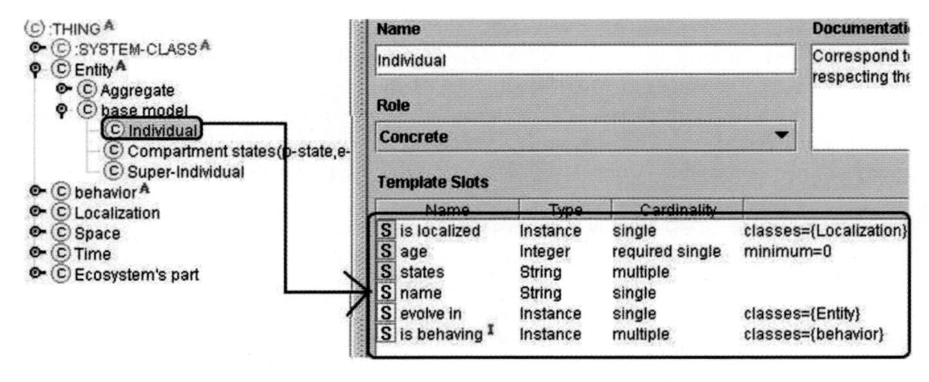

Fig. 1. The individual's class in Protégé

3. System represents a set of entities (or its subclasses) which are interrelated in dynamics. Particular models (as Lotka-Volterra equations or example) exist describing those systems.
4. Compartment stands for a space with particular values describing its abiotic and environmental factors. They embody individuals, global entities and systems.

Those entities are in relation with a behavior standing for the model describing the evolution of its values.

Behaviors are a particular type of classes defining a way to describe the dynamics of an entity. Indeed, it is linking an entity with a particular tool (rule or equation) that will make its values evolve according to the description of space and time. The tools come from mathematics, physics or computer science for example and are classical in modeling. Particular behaviors are associated with the different subclasses of entity and corresponding to their level.

As we sketched previously, entities could be localized in space. Therefore, different type of **space models** have been defined including continuous or discrete ones, network, grid... **Time models** are also available as a set of classes representing some ways to introduce space in a simulation (event driven for example).

Last but not least, a particular class named **ecological's part** or **ecosystem's part**(see figure 2) fulfills a critical task. In fact, this class and those that inherit of it constitute ecological knowledge linked with the modeling concepts from the other parts. Thus, as a user starts an instantiation of the ontology, he applies his ecological knowledge to that part. Then, the links between these parts and the other parts make the user ask himself questions about how to model his case referring to the definition of the ontology. Therefore, the ontology naturally makes him assuming the task of the modeler as he instantiates it.

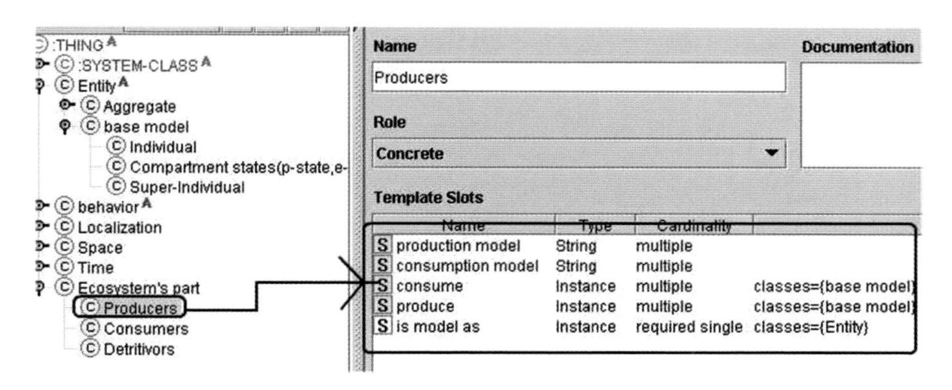

Fig. 2. The producer class : subclass of ecological's part.

Example of three species aquatic ecosystem

The following demonstration corresponds to a down-top approach, but we can also use a top-down one to model our ecosystem. First, the example shows us a food chain composed of four species. As we want to introduce the species, we should first determine if they are consumers, detritivors or producers.

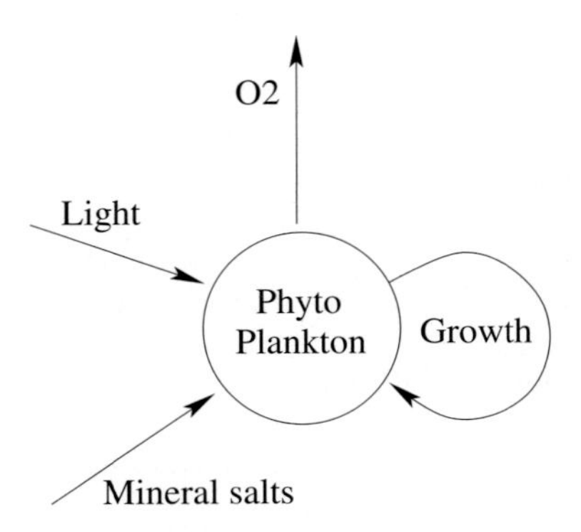

Fig. 3. Producer - phytoplancton

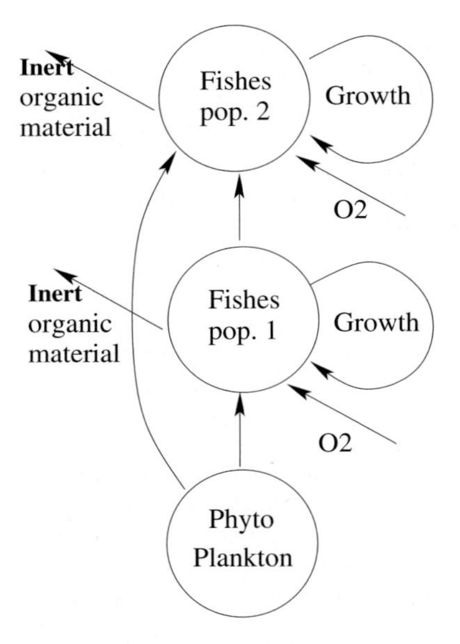

Fig. 4. Consumer - food chain

- **Plankton.** Plankton produces oxygen, consumes mineral salt and multiplies without consuming organic mass. It clearly corresponds to producers (see figure 3).
- **Bacterias.** Bacterias use the organic mass to produce salt through decomposition. They are detritivors (see figure 5).
- **Fishes.** As far as we know, fishes are consumers as they consume organic mass to produce their own (see figure 4).

As we define those species, we see that producers, consumers and detritivors must be linked to a particular individual-based model (individual or super-individual). We should now think about the way we will represent them in our system. As we

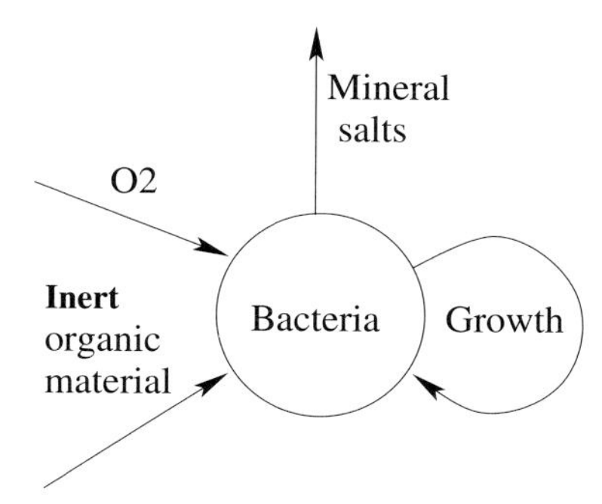

Fig. 5. Decomposer - bacteria

made that choice (individual for the plankton and fishes, super-individual for the bacterias), we must define the behaviors of our individual. Each category has already defined behaviors that can be reused. We can also introduce a new rule-based model that will be used by the behaviors to simulate the photosynthesis for example. Simultaneously, the ontology asks us which way the individuals will be localized in the space. Therefore, we should look for a localization suiting our needs. It will also define the way the space will be modeled. We choose a 2D coordinate localization. The ontology induces that we should choose a space corresponding to that kind of localization. We then choose a 2D continuous space.

The lowest level of our simulation has been achieved. Fortunately, new questions has been revealed by our ontology. Mineral salt, oxygen, light and organic mass needed by the low-level entities have to be modeled. First, should we introduce them as producers, consumers or detritivors? The best way to reintroduce them is to place them at the compartment level as they correspond to e-states. The light is a criterion to distinguish the different compartment. The mineral salt, oxygen and organic mass quantities vary in a compartment depending of its settlement. Therefore we define three e-states. Their behaviors are simple. Oxygen, mineral salt and organic mass do not vary by themselves nor have subtle interactions with other elements of the simulation so they have no behavior. Light could vary depending on the time so we can add a behavior making it lower and higher (actually, we will not).

During the previous step, the ontology led us through the definition of our compartments. They will have four e-states (O^2, mineral salt, light and organic mass). The ontology asks us the types of space model used with that compartment. We have already chosen a 2D continuous space one. It only remains to define the behaviors of our compartment. A compartment exchanges fluxes with its neighbors. So we should define a law-based model corresponding to the fluxes between the different compartments. When defining it, we now realize that we must precise how compartments are linked. Therefore, compartments have a neighborhood and values determining how much they are connected to each neighbor. A compartment has also a localization corresponding to the space model "neighborhood".

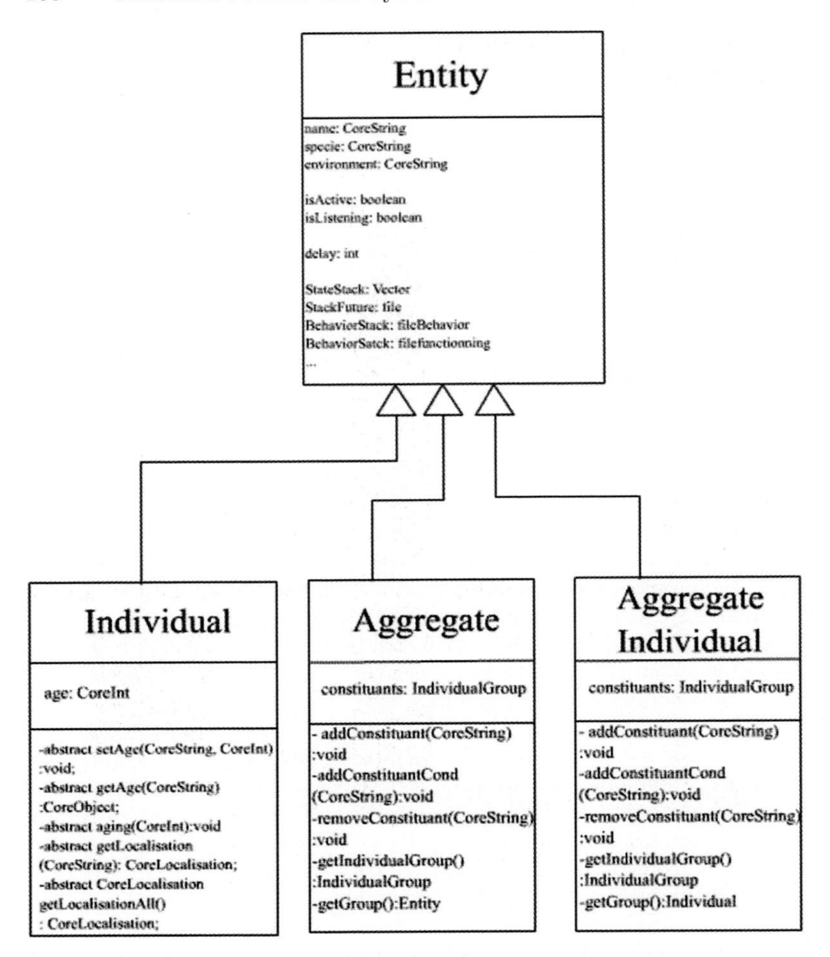

Fig. 6. An example of Class:the class Entity and its child

The informations concerning the neighborhood appear at the level of the ecosystem itself. That one constitutes the higher level of our simulation and embodies the compartments. It also defines the neighborhood as its own space. Its behaviors correspond to the global exchange of the ecosystem with the "outer world" (classical notion in ecology).

To conclude, the ontology naturally guides us through the modeling of a simple example of ecosystem respecting the concepts and assumptions of our own model. That step ends with the choice of classical law-based or rule-based models to simulate the behaviors of each entity. That modeling step achieved, we can now go to the simulation step.

5 From Model to Simulation: a Distributed Implementation

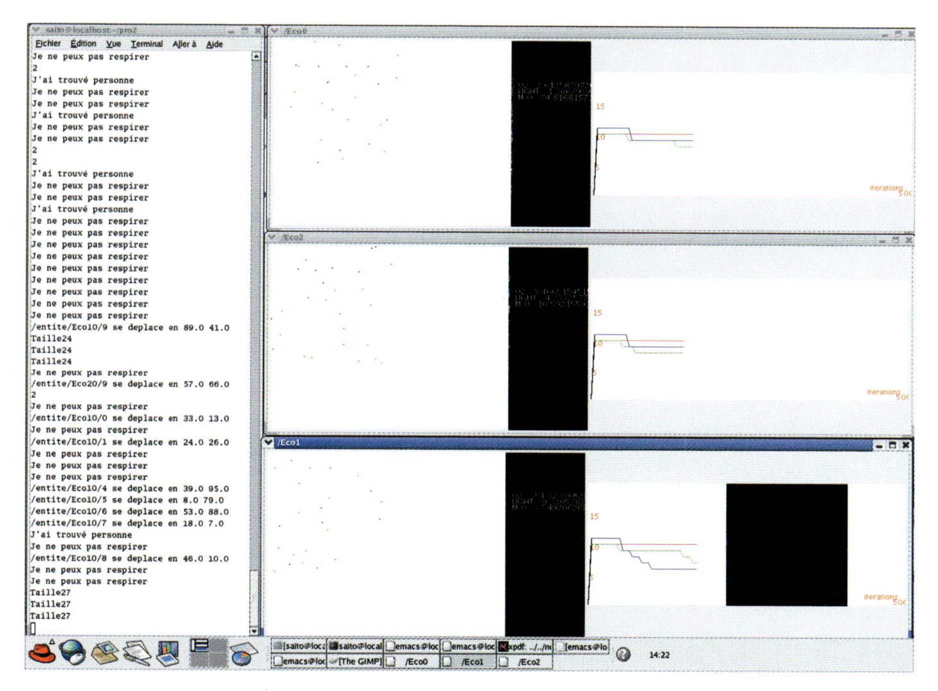

Fig. 7. Screenshot of a simulation

The previous work provides us with a model corresponding to our ecosystem and the aspects we want to study, but the purpose of this model is to give birth to a simulation, allowing us to increase our understanding of the ecosystem. Therefore, we are conceiving a platform corresponding to our model and directly linked with our ontology.

Simulation framework description

First, let us sketch that the framework will follow the concepts of our ontology. We will base the framework on active objects as they suit well the individual-based model chosen for our hybrid model. To easily conceive a simulation tool using active objects, we program in java using ProActive. ProActive [Baude et al. 2000] is an API enabling the conception of distributed active object's applications. ProActive is made as a library of static java classes. The main class (ProActive) allows us to turn Java Object into ProActive Objects (as long as the original object is correctly defined for ProActive). Therefore, ProActive Objects are registered among the network and can communicate asynchronously or migrate. The migration conserves both the object's state and its execution state. Remote Method Call on distant objects are non-blocking. Therefore, we implement all the concepts of our model in a reusable way.

Our framework lays on many classes:

- **Entity.** Every object acting during our simulation is an entity (i.e an active object). It has states, behaviors and a fixed cycle of life. Ideally, each entity have its own execution thread (i.e entities are ProActive Active Objects). In order to maintain a satisfying simulation's speed, special execution objects gather a fixed number of entities and manage their functioning. Even so, entities are active objects and have their own single name (i.e URI) known by all the computers of the network. The choice of gathering entities or not is free and the numbers of entities within an execution object as the number of execution objects on a computer can be fixed by the modeler. This one does not have to manipulate execution objects and only have to define entities.
- **Aggregate (i.e systems).** Aggregates are entity containing other entities. They manage the individuals suppressing dead ones, assuming migration of objects in the network.
- **Compartment.** Compartments are aggregates defining a space and making part of a neighborhood. Compartments have states (e-state or p-state). The states are implemented as Characteristic values which can be active (with their own behavior) or not. In the last case, the compartment's behaviors make them evolve. Compartment behavior's also include flux exchange with other compartments of its neighborhood.
- **Individuals.** Individuals are indecomposable entities localized in a compartment. We distinguish three types of individuals: single individual, super individuals and possibly e-state with behaviors. The individuals have a localization representing its placement within its compartment space. The type (or class and model) of the localization depends on the individual nature.
- **Behavior.** A behavior is a task done by an entity during its life-cycle. Behaviors correspond to the computing application of a model onto an entity. Behaviors can be executed one time, or in a cyclic way. A generic behavior consists in a cyclic call to a method. For example, photosynthesis could be defined as a method in a phytoplankton class. The corresponding behavior would regularly invoke that method depending on the object states and on its local environment. Behaviors corresponding to Automaton are also usable.
- **Time.** Time in the system is divided into two classes. The "event" class represents a dated event in time inducing a change of state or the occurrence of a particular event. The "period" class represents a lap time delimited by two significant events. period can be ecological phase, period of growth ...

  The simulation itself is made to be of discrete time type one for the individual based level. The compartment themselves acts both in an factual way and a discrete way. In fact, fluxes exchanges are made when the compartment has evolved so much that it cannot consider it isolated anymore. Therefore, the fluxes link it with it local neighborhood. Despite that fact, more regular events occur (specially interaction between a compartment and the individuals embodied in it), those events are simulated with discrete time.

- **Space.** A class representing the main characteristic of spacial representation in our simulation. For now, two types of spaces are available. The grid one consists in a traditional 2D grid. A continuous space is also proposed. The system permits the use of multiple space at the same time on condition that one can pass from a space representation to another. Therefore, one can use multiple type of model (individuals, super-individuals). Moreover, Grid induces introduction of a local memory in the space useful to speed up the simulation. Continuous space solve some problems due to local heterogeneity.
- **Localization.** A class representing the main characteristics of localization of an entity in a space. Super Individuals can be localized with:
  - coordinates corresponding to a grid
  - coordinates and a radius defining a circled area corresponding to the Super Individual on a continuous space.

  Individuals are localized with coordinates. As Individuals and Super-Individuals are involved at the same time in simulations, a conversion system makes the different type of localizations compatible. The localization library will be expanded.
- **Living Part.** Those classes represent the main informations needed to incorporate a consumer, a detritivor or a producer in a simulation (inducing the model corresponding to the photosynthesis, decomposition ...). Therefore, when creating an individual or a super-individual, the modeler should define methods the behaviors would call during the simulation.

Many other classes have been developed in addition to those ones that are purely computational, with no meaning in the ontology. Our goal is to provide an easy to reuse and complete framework. Moreover, the final platform will be completed with an interface guiding ecologists through the conception of a simulation without coding.

6 Self regulation of predator-prey systems induced by interactions with environment conditions

Using that model, we can point out drastic feedbacks induced by environment conditions between populations as well as the influence of space on population dynamics.

The aim is to point out the particular population's dynamics induced by the interaction with the environment meaning environment conditions constitute an indirect interaction between populations.

Let us consider two populations :

1. One made of planktons responsible for the photosynthesis and therefore the production of O_2 depending on light intensity. As they do the photosynthesis, planktons raise their organic matter until they can reproduce.
2. One made of fishes consuming O_2 and eating planktons to survive and reproduce.

Those populations are living in a compartment with a particular light intensity in which O_2 is an environmental condition.
We should associate a model with each part of the study:

- Fishes are modeled as super-individuals. Each step of time, a death rate is applied on their number. They also eat preys in their radius of action (depending on their number) which raises their organic matter amount. Eventually, they could reproduce if their organic matter is high enough. If no preys are near, they move until they find one. Finally, they breathe and undergo a loss of number if they cannot.
- Planktons are modeled as super-individuals. Depending on their number and on the light intensity, they produce O_2 and raise their number.
- The compartment has O_2 and light as environment conditions. Light does not vary. O_2 increases periodically due to external supply. The compartment's space is a bidimentional grid.
- Fishes predation rate is calibrated to be far too high compared to planktons reproduction rate.

First, we deprive the fishes of breathing, this meaning they do not need to consume O_2 anymore. Using that partial model, we obtain a well-known case of populations dynamics of type 8. Predator's number undergo a period of increase until prey's number reaches nil. Then, their number falls to nil too as they have no more resource. Preys disappear as predation is too hard. The time for plankton to be annihilated lays on the efficiency of predator to find them all in the space of simulation.

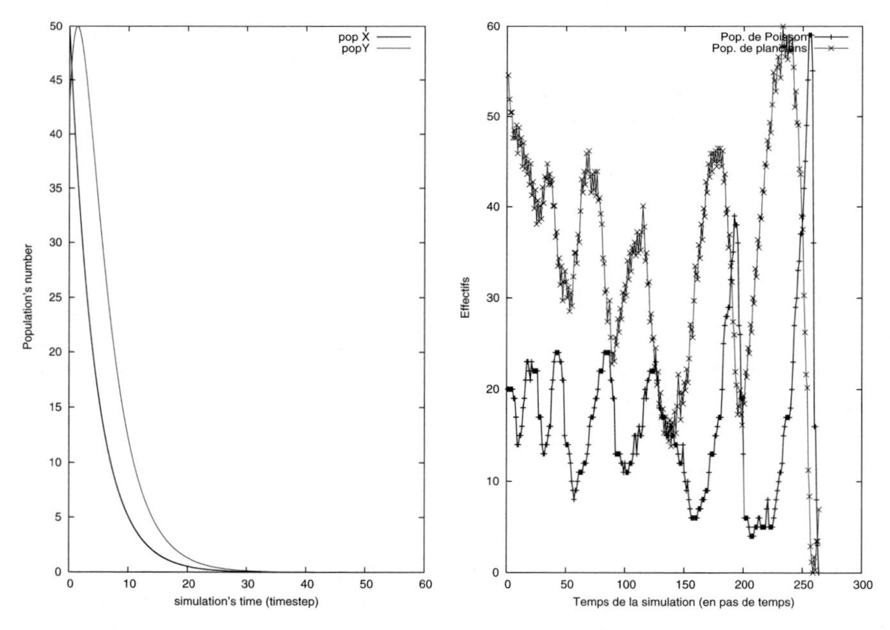

Fig. 8. Simulation curves without breathing activity (on the left part) and with breathing activity (on the right part)

Secondly, we make hypothesis that fishes should breathe to survive. Many simulations show us that predator's number and prey's numbers evolve in phase. First, the number of preys decreases due to predation as the number of fishes increases. As preys became too rare and cannot produce enough O_2 for fishes, fishes population decreases until preys are able to multiply again and cope with the air production. Thus, the number of fishes number can go up again. The previous phenomenon repeats. Eventually, space makes the two population not to encounter. It ends with an excess of O_2 making fishes to eat all planktons without suffering from lack of O_2. A example of simulation trace is shown in figure 8

To conclude, we sketch how environmental conditions constitute a drastic factor of indirect interactions between population dynamics and how the model helps take them into account at several levels.

7 Interaction-based Approach to Detect Systems and Genetic Algorithm to Calibrate the Paramaters

7.1 Analyzing the interaction network of the simulation

As the simulation runs, agents representing individuals or active objects representing populations interact at their own level and with their environment according to their perception. Eventually, those parts become a strong link in dynamics and interactions remains between them constituting a particular interaction network. As we previously stated, we can associate an interaction network to a model. Therefore, the network of an emerging system could correspond to the one of a well-known global model and we can compute the dynamics of that emerging system at a higher level using the corresponding model. Of course, we should be able to define if an emerging system is stable enough to justify a change of scale in our simulation. We simply fix that if a particular interaction network remains stable during a certain period, the emerging system should be reified. Secondly, we should be able to evaluate interactions in order to compare them. To fulfill this task, we consider the influence on the organic mass of the elements in interaction with another element induced by the interactions during the significant period. This number is the valuation of the edge between the nodes representing the two elements and pointing to the node representing the predator.

Figure 9 shows two cases of interaction networks corresponding to our case of study described in 6. First one simply implies that individuals and super-individuals are in interaction with the same individuals and thus can be consider as a a single super-individual. Second one corresponds to two sets of individuals (preys and predators) that only interact together meaning the set of predators only feed on the set of preys and the set of preys is only consumed by the set of predators. Therefore, they stand for a local system that can be modelled thanks to a Lotka-Volterra model linking the dynamics of the two populations. By the way, we could extend those interaction networks to the general case as shown in figure 10.

Thus, the simulation dynamically achieves the making of the interaction network corresponding to the computation of its model on the lower level (IBM). In parallel,

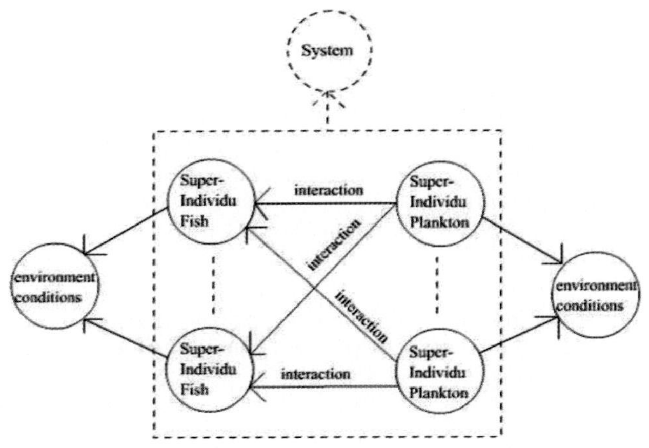

Individual interaction network corresponding to a Lotka-Volterra system

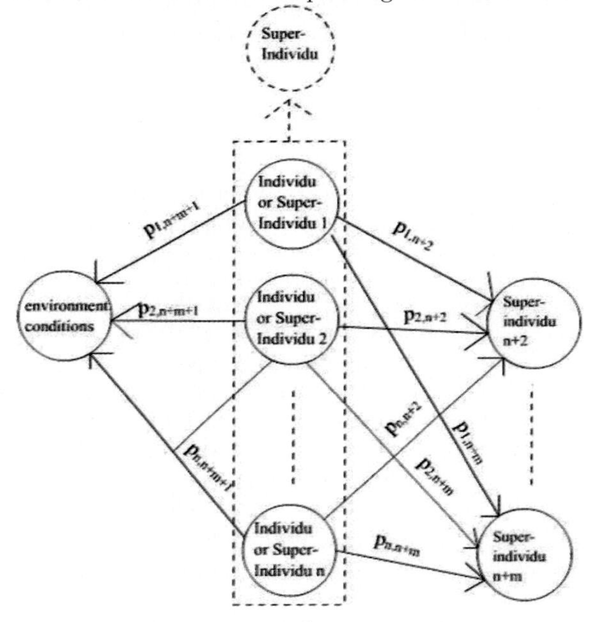

$$\forall x1, x2 \in [1,n], \forall y \in [n+1, n+m+1], |p_{x1,y} - p_{x2,y}| < Beta$$

Individual interaction network corresponding to a super-individual

Fig. 9. Individuals interaction network according to the model

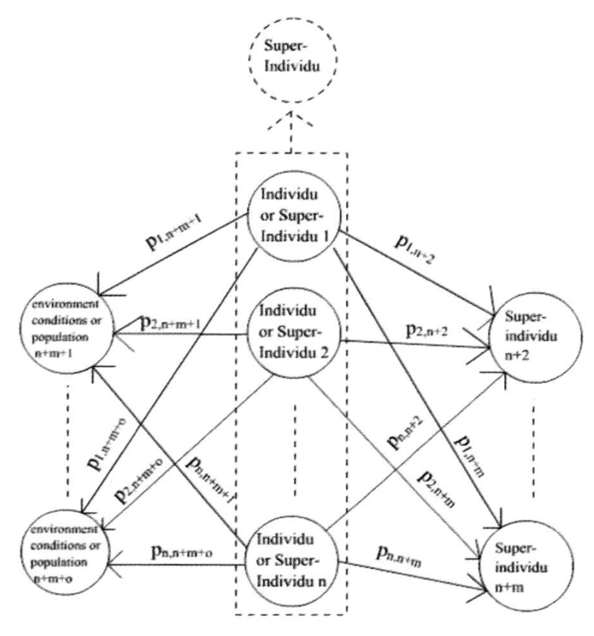

Fig. 10. General individual interaction network corresponding to a Lotka-Volterra system

it analyzes this interaction network in order to detect local interactions networks corresponding to a particular global model. When a local interaction network remains stable for a sufficient period of time, the simulation gathers its elements in a system corresponding to the associated global model and computes it on the global level.

7.2 From IBM to global model: calibration with genetic algorithm

Passing from discrete model to continuous one

We see how we can pass from a set of individuals on the IBM level to a global system on the higher level. One of the critical problem is that IBM and global models can differ in nature. In our example, IBM are discrete models whereas global models could be continuous. It is the case when passing from a set of individual to a Lotka-Volterra model. In fact, the number of individuals of the set of super-individuals evolve according to a discrete time but we must recreate the same dynamic with a Lotka-Volterra model.

$$\begin{cases} \frac{dX}{dt} = a.X - b.X.Y \\ \\ \frac{dY}{dt} = -c.Y + d.X.Y \end{cases}$$

where a, b, c, d are positive values belonging to $[0, 1]$.
The population dynamics between time t_0 and time t_n of the set of individuals on the IBM level is a discrete set of X_i, Y_i where X_i is the number of preys and Y_i the number of predators with $i \in [t_0,t_n]$. Therefore, we will try to determinate parameters a,b,c and d for the Lotka-Volterra equations ideally to verify $X(i) = X_i$ and $Y(i) = Y_i$ with i $\in [t_0,t_n]$. Of course, we will mostly search for solutions nearly matching that criteria using an heuristic approach.

The genetic algorithm

Genetic algorithms belong to meta-heuristic approaches for finding approximate solutions to optimization problems. The principle relies on the crossing and mutation of a population of chromosomes representing solutions of our problem. An heuristic is used to evaluate the solutions. Using that fitness, a selection is made to keep and cross over the best solutions in order to generate a new population that will serve in a new cycle of the algorithm. The genetic algorithm goes on until a satisfying solution has been found.

Let us consider a population X of preys and a population Y of predators and n+1 points $(X_i,\, Y_i)$ with i $\in [t_0,t_n]$ representing the trace of their numbers coming from their simulation at the individual level from time t_0 to t_n. Our concern is to fix parameters a,b,c and d for the non-linear system described in 7.2 to minimize the least square criteria:

$$\sum_{i=0}^{n} (X(t_i) - X_i)^2 + (Y(t_i) - Y_i)^2$$

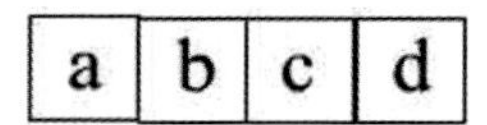

Fig. 11. Chromosome used in the algorithm.

That fitness allows us to find solutions matching the phase of the set of points. Another issue has to be taken into account. In fact, the space of solutions is clearly wide as the algorithm must fix the four parameters whose values belong to the continuous interval $[0, 1]$. So, the population chromosome of the genetic algorithm used here, is the sequence of the parameters (a, b, c, d) as described in the figure 11. Thus, the initial population of chromosomes only represents a few part of all the possible combined solutions. Therefore, we add to the mutation another factor which introduce new genes : migration. That means that every new worse solutions of the populations are replaced by new chromosomes. Those arrival corresponds to the mechanism induced by migration between populations.

Large scale tests have been made to check many features. We wanted to know whether the migration mechanism improves the algorithm or not and if the better solution comes from a crossover or not. Finally, we examine the influence of the number of chromosomes and loops on the efficiency of the algorithm.

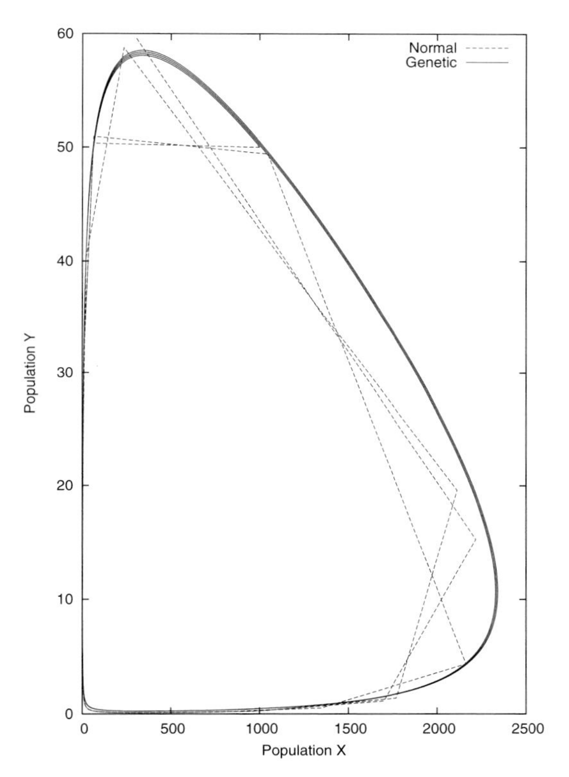

Fig. 12. Phase diagram comparison between the set of points from the IBM simulation (labeled Normal) and the deduced Lotka-Volterra system computed by genetic algorithm (labeled Genetic).

We end with the following conclusions:

1. all of the solutions of the algorithm are produced by a crossover ;
2. with number of chromosomes and loops fixed, near the totality of the solutions were far better with the modified algorithm (using migration);
3. the greatest the population is, the fastest the algorithm improves its solution and the better it is to avoid local extrema ;
4. the number of loops has an influence on the performance of the algorithm as more loops implies a better solution. At the first time, the improvement is drastic until it reaches a critical number of loops where it is hard to get a better

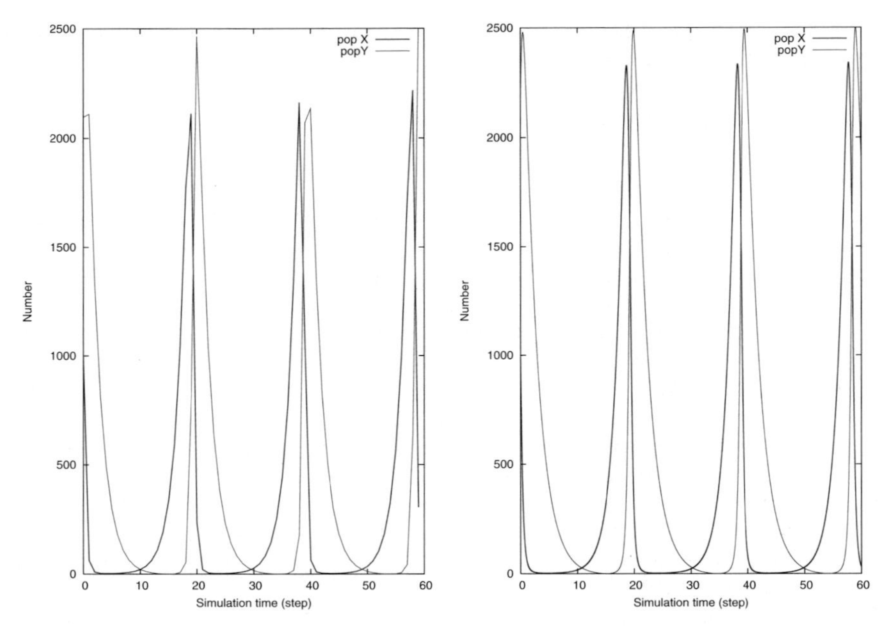

Fig. 13. Population evolution from IBM simulation (on the left) and from the deduced Lotka-Volterra system (on the right)

solution. So, the curve linking the number of loop and the fitness of the solution looks like a logarithmic model one.

Figure 12 shows the difference of phase between the set of points from IBM simulation and the deduced Lotka-Volterra system computed by genetic algorithm (on figure 13).

8 Conclusion

The present works focuse on studying natural systems and especially the way their complexity expresses in their dynamics. We mainly try to integrate previous work and restore them in the field of complexity. By choosing that approach, we adopt a bottom-up point of view. In fact, we base our work on the analysis of the becoming of elements of systems and mainly on their interactions. Investigating population dynamics, we exploit the individual based models results and interrelate those with global modeling to study natural systems structure dynamics and the link between abiotic factors and populations. The end of the paper presents a way to take advantage of previous works in modeling by adopting the interaction-based point of view to appreciate the nature of the systems we detect. As we do so, we are able to give a relevant mathematical expression of the system as we see with the example of the Lotka-Volterra systems and the genetic algorithms. This technique should be generalized to other kinds of systems and different global models. The whole work is a contribution to the analysis of IBM results in the field of complex systems and constitute a link between global modeling and IBM.

References

[DeAngelis et al. 1992] DeAngelis, D.; and L. Gross, editors (1992) *Individual-based models and approches in ecology.* Chapman and Hall.

[Baude et al. 2000] Baude, F.; Caromel, D.; Huet, F. and J. Vayssier (2000) *Communicating mobile active object in Java.* in Proceeding of HPCN Europe.

[Von Bertalanffy 1968] Von Bertalanffy, L.. (1968) *General system theory foundations development applications.* George Brazille Inc., New York.

[Cardon 2004] Cardon, A. (2004) *Modéliser et concevoir une machine pensante.* Vuibert - collection Automates Intelligents.

[Cardon 2005] Cardon, A. (2005) *La complexité organisée, systèmes adaptatifs et champ organisationnel.* Hermès-Lavoisier.

[Frontier 1999] Frontier, S. (1999) *Les écosystèmes.* PUF.

[Grim 1999] Grim, V. (1999) Ten years of individual-based modelling in ecology: what have we learned and what could we learn in the furture. *Ecological Modelling*, (115):129–148.

[Gruber 1991] Gruber, T. (1991) *The role of common ontology in achieving sharable, reusable knowledge bases.* in proceeding of KR'91, Principles of Knowledge Representation and Reasoning: second international conference.

[Koestler et al. 1969] Koestler, A.; and J. Smythies (1969) *Beyond reductionism.* Hutchinson.

[Le Moigne 1994] Le Moigne J.-L. (1994) *Modélisation des systèmes complexes.* Dunod.

[Prévost 2005] Prévost, G. (2005) *Modélisation d'écosystème multi-niveaux par des approches mixtes.* PhD thesis, University of Le Havre.

[Protégé 2008] Protégé Ontology Editor and Knowledge Acquisition System (2008) `http://protege.stanford.edu`.

[Scheffer et al. 1995] Scheffer, M.; Baveco, J.M.; DeAngelis, D.L.; Rose, K.A. and E.H. van Nes (1995) *Super-individuals a simple solution for modelling large populations on individual basis.* Ecological Modelling 80, 161-170.

[Tansley 1935] Tansley, A. (1935) *The use and abuse of vegetational concepts and terms.* Ecology, 16(3): 299.

Part IV

Emotion and cognition modelling

Simulation of Emotional Processes in Decision Making

Karim Mahboub and Véronique Jay

LITIS, University of Le Havre
25, rue Philippe Lebon, B.P. 540
76058 Le Havre Cedex, France
{Karim.Mahboub, Veronique.Jay}@litislab.eu

Summary. Human emotional capabilities have recently been considered essential in decision making. An emotional model applied to the *Gambling Task* game is outlined here. The aim is to be able to simulate a human behavior with respect to the emotional feedback created by the environment. To do so, an OCC (Ortony, Clore and Collins) model emotion type is used to define several criteria representing a human emotional structure. Moreover, a probabilistic graph is brought into play for the behavior representation through the game environment. Results are promising and show that the emotional reactivity is coherent and globally lead the player to his objectives.

Key words: emotion, decision making, gambling task, behavioral graph

1 Introduction

This study was brought to life from the assumption that emotion could be part of human intelligence and particularly be of a great help in decision making. Indeed, Damasio's works [Bechara et al. 2000] have proved that with a lack of emotional activity, a human being is hardly able to make a reasonable decision when facing a problem.

The objective of this work is to simulate emotions using a sociological approach in a very particular context in the world of cognitive psychology: the *Gambling Task*. To do so, we worked together with the *Psychology and cognitive neurosciences laboratory* of Rouen (France). This collaboration will allow us to draw a parallel between human behaviors and computed simulation outcomes.

After having a look at the sociological aspects of emotions, we will develop a model in response to the problem of emotions simulation. Finally, the results obtained and the planned perspectives will be presented.

2 State of the art

When looking at the different definitions of emotion on the internet or in any kind of book, we realise how difficult it is to comprehend all the aspects involved in emotional processes. Hence, several sociologists and psychologists have tried to make a list of all the possible human emotions. Most of them have finally determined a few basic emotions that are considered being sufficient to represent any human feeling. For instance, Ekman, a well-known American psychologist, describes 6 *basic* emotions (also called *primary* emotions): *anger, disgust, fear, joy, sadness,* and *surprise.*

Apart from the actual distinction of these emotional concepts, psychologists, from their point of view, tried to make different types of categorisations for emotions. One of the most famous is the *OCC model*, made by three psycho-cogniticians: Ortony, Clore and Collins [Ortony et al.].
According to these three authors, emotions come from the appraisal of three different aspects of the world: the consequences of events, the actions of agents and the perception of an object. For instance, an event which aims at realising a particular objective will create joy; an action from an agent (individual) that would go against his principles will end up with a feeling of shame; the perception of an object can be disgusting depending on the agent preferences, etc.

		+	-
Consequences of events	For others	Happy for	Resentment
		Gloating	Pity
	For self	Hope	Fear
		Joy	Distress
Action of Agents	Self Agent	Pride	Shame
		Gratification	Remorse
		Gratitude	Anger
	Other Agent	Admiration	Reproach
		Gratification	Remorse
		Gratitude	Anger
Aspects of Objects		Love	Hate

Fig. 1. The OCC Model

Therefore, the OCC model (see figure 1) defines three classes of emotions, according to the emotional context they refer to, in which we have smaller groups concerning the person responsible for the triggering of the action (generally 'other' and 'oneself'). Each emotional dimension is represented by an antagonistic couple, like *joy* and *distress*, or *love* and *hate*, in which an individual's emotional state is located, somewhere between the two bounds.

3 Model of emotion

Within the framework of this project, the modelisation which is used is based on the OCC system described above. For each antagonistic couple previously listed, we associate a sigmoid function called μ, defined on the interval $[-1;1]$, as follows:

$$\mu : [-1;1] \mapsto [0;1]$$
$$k \rightarrow \mu(k)$$

The variable named k represents the current position of the emotional criterion on the curve. According to its value, the actual emotional criterion is, more or less, positive or negative. Therefore, if k equals 0, the emotional state related to the couple *joy-distress* is neutral. The more k is close to -1, the more important the feeling of distress, and vice versa, the more k is close to 1, the more important the feeling of joy.

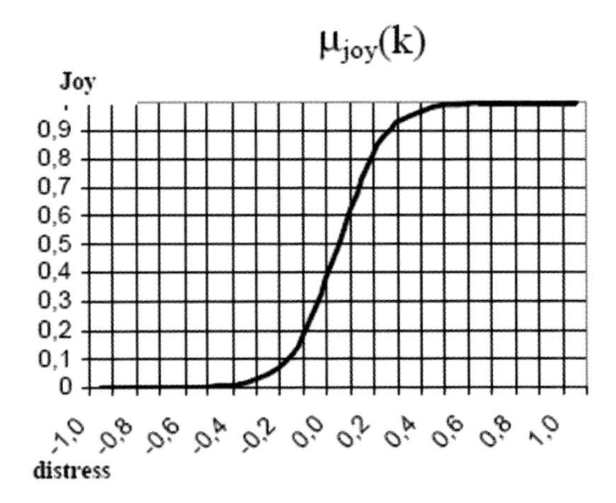

Fig. 2. The μ_{joy} function

The shape of this sigmoid function can vary, with respect to the feeling couple which is modelised [Colloc et al. 2004]. However, it can also change depending on the actual individual involved. As a result, an easily depressed person will have a very steep *joy-distress* curve, with the ability to go quickly from a great joy to a deep sorrow. On the other hand, a mentally strong person will end up with a gentle curve.

Here is the sigmoid function used in the model:

$$\mu(k) = \frac{1}{(1 + e^{-\lambda k})}$$

The λ value is proportional to the curve slope.

Gambling Task

The Gambling Task (see Bechara and Damasio's works [Bechara et al. 2000]) is a well-known test which aims at stimulating the emotional processes of a player by giving him some money and proposing him to increase his capital through a card game.

This game consists of four decks of cards. On each card a number representing a certain amount of money, positive or negative, is written. For each card taken, the player wins or looses money with respect to the amount indicated on the card. The game stops when a hundred cards have been taken.

Of course, in order to make the game more interesting, the player is told that amongst the four packets, two of them are more profitable than the other two. Indeed, the average sum of all the cards located in a profitable packet is positive, meaning that the player has more chance to earn money. However, the sums written on these cards are not high. Eventually, the profitable packets make the player earn small amounts of money without loosing much. On the other hand, the bad packets are not profitable at all, and make the player loose a lot, even more than he will ever earn on these packets.

Finally, the player's goal is to collect money by identifying the good packets. He will then unconsciously develop a strategy to achieve this goal.

Probabilistic behavioral graph

During the game, the player obeys a behavioral graph that describes the environmental protocol. In other words, each packet is represented by a state in the graph, and a connection between two states is a choice that the player made to go from a packet to another.

In the following example (figure 3), the player has just taken a card from deck 1 and is planning to choose the next card. He has four possible choices: staying in packet 1 or changing for packet 2, 3 or 4.

Each possible choice is linked with a particular probability p, calculated from the previous results obtained in the four decks. Hence, the sum of the four probabilities stemming from the first deck must be 1.

The p_{En} are the entry values for each state (used only to choose the first card of the game).

Decision making process

Decision making consists in :

1. Defining the values in the behavioral graph according to the player emotional state and the history of every deck of cards. To do so, we need to calculate every P value in the graph, knowing the emotional $\mu_i(k_i)$ curves (i being an emotional couple), and using valence function α_i as follows :

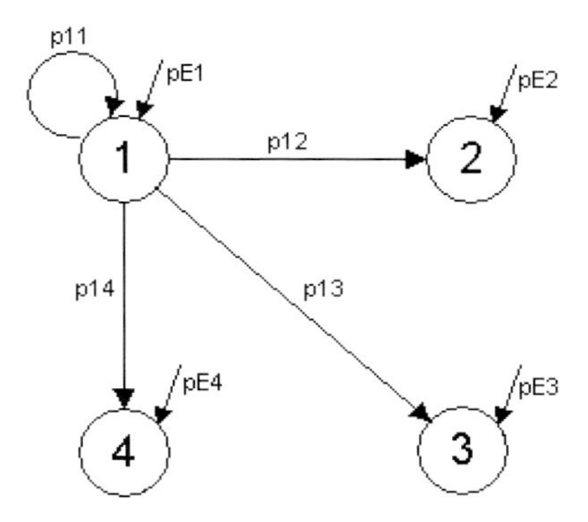

Fig. 3. The behavioral graph

$$\forall transition\ t\ \forall criterion\ i \begin{cases} p_i = \alpha_i(t)\mu_i(k_i) \\ P_t = \sum p_i \end{cases}$$

2. Calculating the score of every deck, using the gains and losses that occurred so far (the history of each deck). The default equation used is :

$$Score = \frac{(gains \times gainMax) - (losses \times lossMax)}{(gains \times gainMax) + (losses \times lossMax)}$$

   The *gains* and *losses* values correspond to the sum of all the positive and negative cards that have been taken on the deck, $gainMax$ and $lossMax$ being the best and the worst card taken. When calculated, the $score_i$ value is then multiplied by the probability value P_i, in order to influence the latter positively or negatively.
3. Normalising the P_i values, so that their sum equals 1.
4. Choosing randomly a packet in which the player is going to take a card, according to the P_i values.

Emotional feedback

Emotional feedback is the effect of decision over the mind. In the Gambling Task, this phenomenon depends on the game situation, described by the last card taken. The influence of the card over the player's emotion is composed of different parameters.

Firstly, we assume that the intensity of the emotion is directly proportional to the value of the card taken, positively or negatively.

Secondly, we need to take into account the capital of the player at the time he takes the card. Indeed, we can easily consider that a card indicating a strong loss will not especially have a big influence if the player has a lot of money at that time.

Finally, we consider that the emotional state of a normal human being can not be strongly modified, in a realistic manner. Therefore, the emotional feedback will not be computed directly through an equation but using a differential modification of the concerned parameters, with the idea that the emotional state is to be changed gradually, without brutal transformation.

Considering all these assumptions, we obtain the following equation:

$$k_i = k_i + \frac{v}{c} \times r \times \alpha_i$$

i : index of the emotional criterion;
k_i : value of the emotional criterion curve on the X axis;
v : value of the card taken;
c : player capital just before the draw;
r : "mind resistance" coefficient ($0 \leq r \leq 1$);
α_i : magnitude of the emotional criterion compared with the other criteria.

The r coefficient represents the player emotionalism, defined in the interval $[0; 1]$. It is his ability to keep his calm in high-rate emotional situations. The more this value is closed to 1 and the more sensitive the player is, and vice versa.

4 Implementation and results

ModEm is an application that directly implements the emotional model previously seen. It draws a probabilistic graph representing the Gambling Task environment and shows all the results related to the emotional state of the player as well as his situation in terms of money and decision parameters. Moreover, it uses a special file format which describes the entire game protocol:

- The global parameters: number of turns and starting capital;
- The composition of the packets: number of packets, size and description of all the cards;
- The emotional system: emotional model used, emotional resistance, λ values for each curve, and starting k values.

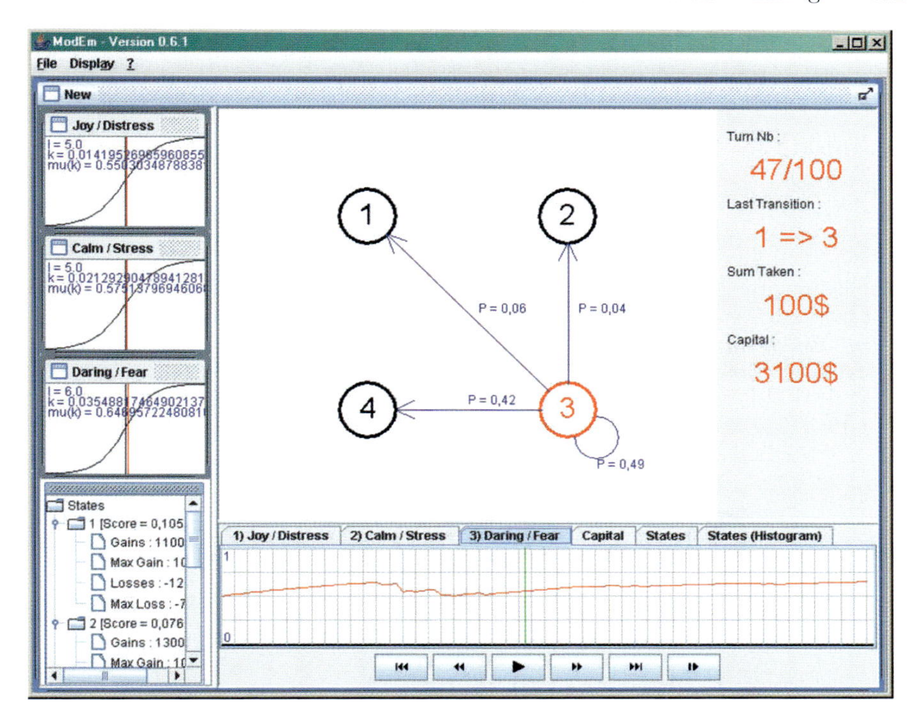

Fig. 4. The graphical user interface

```
************************************
#ModEm - Version 0.6.1
************************************

#TURN_COUNT 100
#CAPITAL 2000
#PACKET_SIZE 40
#PACKET_COUNT 4
#PACKET_CARDS 100 100 -500 ...
#PACKET_CARDS 100 100 -1250 ...
#PACKET_CARDS 50 50 -150 ...
#PACKET_CARDS 50 50 -250 ...
#FORCE_STATE 0 (1)
[#FORCED_STATES 0 2 4 3 1 ...]

#EMOTIONAL_MODEL KM
#RESISTANCE 0.5
#LAMBDA_VALUES 5.0 5.0 6.0
#K_VALUES 0.0 0.0 0.0

#ALPHA_VALUES 0.2 0.3 0.5
#PROBA_VALUES 1.0 1.0 1.0
#PROBA_CHOICE 1
#SCORE_CHOICE 1
```

Finally, the application is able to store its information, with the aim to allow the user to reach any previous played turn in the same game. It can also force the player to choose a particular packet for each turn by using a specified sequence of packet numbers written in the file.

Results

In this part, we will have a look at the obtained curves and graphs in order to analyse the global coherence of the application.

The behavioral graph initially describes four states corresponding to the four decks of cards, each state being valued 0.25, i.e. 25%. This is coherent since the player, before taking his first card, has no information about the decks. Therefore, their probabilities are equal.

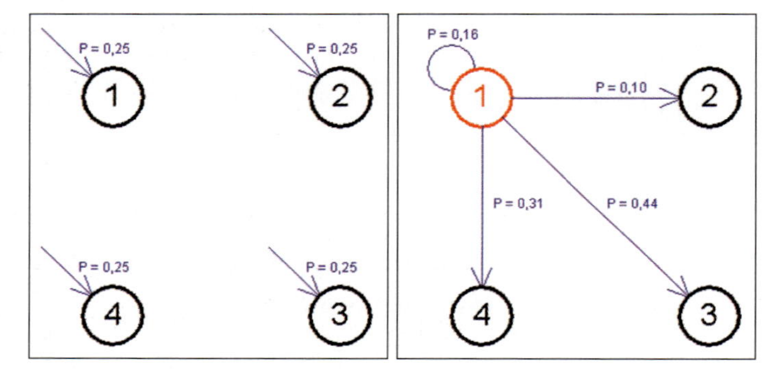

Fig. 5. Example of game graph (start and end)

At the end of the simulation, probabilities change, allowing the game to distinguish the beneficial packets from the unfavorable ones. Hence, in figure 5, we clearly see that decks 1 and 2 (the disadvantageous packets) have respectively 16% and 10% of chance to be selected, wether decks 3 and 4 (the advantageous packets) have 44% and 31%.

The application also produces behavioral curves that give information about the evolution of the game and the emotional state of the player.

Emotional curves (see figure 6) are defined between 0 and 1 and show the different values of the $\mu(k)$ function according to time.

The capital curve (see figure 7) helps to see the evolution of the capital throughout the game.

Finally, the state curve (see figure 8) gives details about the different choices of packets. It can either be seen as a curve, or in the form of a histogram.

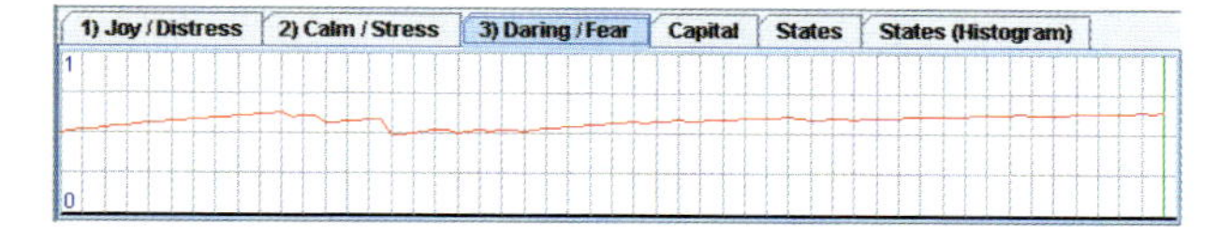

Fig. 6. Emotional curve example ($\mu(k)$)

Fig. 7. Capital curve example

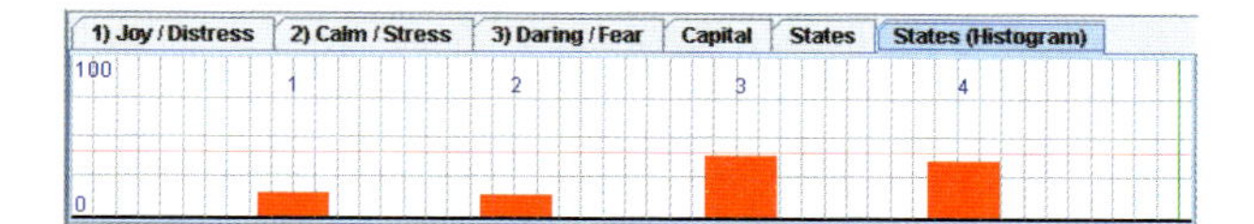

Fig. 8. State histogram example

We can notice that the player has taken much more cards in packets 3 and 4 (the good ones), which means that he is naturally more attracted by these packets. This behavior is rather close to reality since a normal human being who plays the Gambling Task usually react that way.

5 Conclusion and prospects

In order to improve the player environmental comprehension, we aim to complete the model structure using cognitive agents. For a better adaptation to more important problems, we need to add a cognitive module that will be responsible for the player reasoning abilities and also memory issues.

In order to simulate cognitive capabilities, we use Cognitive Maps (see [Axelrod 1976]). These maps are an adaptive manner of learning new knowledge from the environment by adding new states in the behavioral graph. Thus, this cognitive approach allows the player to develop his own experience through the events stored in his knowledge base representing his memory. The emotion is then considered as a base layer for the decision making procedure, synchronised with the cognitive module. This evolutionary model can not only allow a growth of intelligence through experience, but also communication between agents, with the aim to share knowledge, feelings or points of view.

A second aspect which allows a better understanding is the validation of the application with the help of the experiments made in the *PSY.CO* laboratory in Rouen

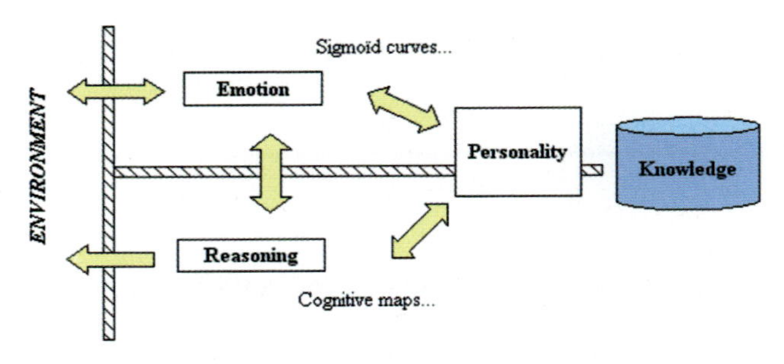

Fig. 9. Mental representation of an emotional cognitive agent

by comparing human nervous intensity measures with emotional behavior curves obtained within the program.

The OCC model, one of the most famous emotional representation created in the domain of sociology, has been successfully implemented in a computer simulation. The example of decision making through the Gambling Task context shows that emotion is a wonderful driving force which strongly leads the player with his choices. Results are consistent and globally correspond to a real human behavior when making a decision involving stress through the gain or loss of money.

However, emotion is a very complex entity, with many different parameters to take into account. So far, we used the OCC model to simulate emotion, but this model only consists of a list of basic emotions in a sociological point of view. There is no deep understanding of the origin and the evolution of emotional processes, and sociology usually has a meta-comprehension of human cognitive activity. In order to create an accurate simulation of human emotion, we need to take care of neurological issues and developmental psychology approaches. This is probably the future of emotion simulation.

References

[Axelrod 1976] Axelrod, R. (1976) *Structure of decision.* Princeton University Press, Princeton, New Jersey.

[Bechara et al. 2000] Bechara, A.; Damasio, H. and A.R. Damasio (2000) *Emotion, decision making and the orbitofrontal cortex.* Cerebral Cortex, 10:295-307.

[Colloc et al. 2004] Colloc, J. and C. Bertelle (2004) *Multilayer Agent-Based Model for Decision Support System Using Psychological Structure and Emotional States.* Proceedings of ESM'2004. Unesco, Paris, France.

[Derre 2004] Derre, M. (2004) *Modélisation à l'aide d'une approche multi-agents de l'émotion, de la structure psychique, de la pathologie dans la prise de décision.* MSc Thesis, Le Havre, France.

[Damasio 1995] Damasio, A.R. (1995) *Descartes' Error: Emotion, reason and the human brain.* Harper Perennial.

[Damasio 2003] Damasio, A.R. (2003) *Spinoza avait raison - Joie et tristesse, le cerveau des émotions*. Odile Jacob.

[Ferber 1995] Ferber, J. (1995) *Les systèmes multi-agents - Vers une intelligence collective*. InterEditions, Paris, France.

[Humaine 2008] The HUMAINE Website (Human-Machine Interaction Network on Emotion) (2008)
`http://emotion-research.net/`

[Obernesser 2003] Obernesser, C. (2003) *Cartes cognitives pour la modélisation comportementale*. MSc Thesis, Bordeaux, France.

[Ortony et al.] Ortony, A.; Clore, G.L.; and A. Collins (1988) *The cognitive structure of emotions*. Cambridge University Press, Cambridge, MA.

[Ortony et al. 1990] Ortony, A.; and T.J. Turner (1990) *What's basic about basic emotions?*. Psychological Review, 97:315-331.

[Plutchik 1980] Plutchik, R. (1980) *A general psychoevolutionary theory of emotion*. Emotion: Theory, research, and experience, 1:3-33.

[Roseman et al.] Roseman, I.J.; Jose, P.E. and M.S. Spindel (1990) *Appraisals of emotion-eliciting events: testing a theory of discrete emotions*. Journal of Personality and Social Psychology, 59:899-915.

[Sloman 2005] Sloman, A. (2005) *Do machines, natural or artificial, really need emotions?*. NWO Cognition Programme, Utrecht, 24th june 2005,
`http://www.cs.bham.ac.uk/research/cogaff/talks/cafe-emotions-machines.pdf`

[Velàsquez 1997] Velàsquez, J.D. (1997) *Modeling emotions and other motivations in synthetic agents*. In Fourteenth national conference on artificial intelligence and ninth innovative applications of artificial intelligence conference. Menlo Park

Emotions: Theoretical Models and Clinical Implications

Sophie Baudic[1] and Gérard H. E. Duchamp[2]

[1] Inserm U 792
Physiopathologie et Pharmacologie Clinique de la Douleur
Hôpital Ambroise Paré
9 av Charles de Gaulle
F-92104 Boulogne Cedex
France
Sophie.baudic@inserm.fr

[2] Laboratoire d'Informatique de Paris Nord - UMR CNRS 7030
Institut Galilée - Université Paris Nord
99 av Jean Baptiste Clément
F-93430 Villetaneuse
France
ghed@lipn.univ-paris13.fr

Key words: Psychology, Biology, Neural network.

Summary. The confrontation between theory and clinical practice contributes to improve our knowledge of emotions. In the present paper, theoretical models both cognitive and biological are briefly reviewed in the first part. Interactions between research and clinical practice are examined across two separate approaches in the second part. The first, neuropsychological approach considers Alzheimer's disease to further explore relationships between emotions and memory disturbance. The second considers panic disorder to explore cognitive and behaviour theories of fear. Finally, we address therapeutic applications for each approach which come directly from the theory.

1 Introduction

The goal of this paper is to provide a concise review of theoretical models and clinical implications in the domain of emotions for the use of Computer Scientists of Complexity Theory. By means of this review, we hope, firstly, to introduce a new interdisciplinary subject and secondly to initiate collaborations from which we expect results both in the field of Therapeutic practice and also in Computer research.

The interaction between researchers and clinicians is essential for the evolution of the discipline and for patients. Theoretical models provide insights onto the structural brain and functional organization of human emotions and therefore their evolution is crucial for the management of patients. The therapeutic actions, the strategies of rehabilitation and the development of new tools are based on theory. This paper depicts in the first part cognitive and biological models and it considers in the second part two approaches which illustrate the implication of theory in the management of patients. The first is that of neuropsychology which analyses the dysfunctions caused by Alzheimer's disease to emotions and the second is that of cognitive and behaviour theories applied by therapists in the management of panic disorder which is a good model for studying fear in humans.

2 Theoretical Models

The study of emotions is hampered by several conceptual problems. The major one is the relationship between emotions and cognition which remains very controversial in the debate between biological [Zajonc 1980] and cognitive [Lazarus 1982] theorists of emotions. The former maintains that emotions and cognition are two independent systems whereas the latter argues that cognition plays an integral role in emotions (for an overview see Schorr [Schorr 2001]). Before presenting the different theories of emotions, it essential to define them, because there is a tendency to group together a number of different affective phenomena under the term of emotions.

2.1 How to define emotions?

The exact nature of emotions remains controversial. Disagreements mainly stem from the fact that the sets of phenomena taken into account are very different. Some focus their attention on the simplest aspects of emotions as they appear in animals or in the early stages of the human development. Others are attracted by the complexity of emotional phenomena. Emotions have two dimensions: expressive and cognitive. The expressive dimension concerns the production of facial emotions or expression of internal states of the brain whereas the cognitive dimension is related to the comprehension of facial emotions expressions and intentions to act. Emotions are feelings that accompany emergent "states of being" like happiness or despair. According to Ekman [Ekman 1984], emotions are reactions that last several seconds. They must be distinguished from very brief responses such as the reflex reactions or basic survival-related appetitive behaviours and from long-lasting affective schemata, such as affects or personality traits. According to Scherer [Scherer 1984], emotions are different from a simple state of the organism (behaviourists). They are processes, i.e. a dynamic sequence of different variables whose components are the following: cognitive evaluation of stimuli or situation, physiological activation, motor and facial expression, action draft or planning of behaviour and feelings (subjective states). In accordance with this point of view, the function of emotions is adaptive because it allows a large flexibility of behaviour thanks to an elaborate treatment of complex information. Examples of emotions are happiness, sadness, anger, fear, surprise, disgust and also despair, pride... The next part of the present topic provides an overview of major theories of emotions. It summarizes and combines several authors' works: Gainotti [Gainotti 2001], Scherer and Peper [Scherer et al. 2001].

2.2 Cognitive models of emotions

There are two major models; the componential and the hierarchical. See Scherer and Peper [Scherer et al. 2001] for an excellent review on emotion theories.

Componential models

Emotions have qualitatively different facets. According to Scherer and Peper [Scherer et al. 2001], most of the psychological theories postulate that subjective experience, peripheral physiological responses and motor expression are major components of the emotion construct. These three components have often been called the "*emotional response triad*". A few theorists include two other components: cognitive and motivational in the emotion process. In Scherer and Peper's view [Scherer et al. 2001], componential models differ strongly with respect to the relative role assigned or the amount of attention paid to these different components. Systematic researches in this field are associated with Lazarus [Lazarus 1982] and Scherer [Scherer 1984].

Hierarchical models

In Gainotti's view [Gainotti 2001], we can distinguish within this classification "structural from developmental" models. The *structural models* maintain that emotions are hierarchically organized with numerous discrete emotions at a basic level and very few emotional dimensions at a higher level. The basic emotions such as happiness, sadness or fear are viewed as the building blocks of the emotion system while the dimensions of valence, (pleasantness/unpleasantness) and arousal (rest or activation) take place on a higher level in the hierarchy. Emotions at the basic level have an important adaptive function and are devoted to the detection of stimuli that are crucial for the organism's well-being. They appear at the earliest stages of child development. Existence of differences between basic and complex emotions is at the origin of the construction of developmental models. The partisans of these models admit that complex emotions derive from the basic ones. According to Gainotti [Gainotti 2001], functioning of these *developmental models* is based on the activity of three functional levels: the sensorimotor, the schematic and the conceptual. The *sensorimotor* level is at the first grade of action or interpersonal communication, the basic interactive schemata of human species. It involves a set of expressive-motor programs that are innate and universal. The *schematic level* which is the second stage of emotional processing includes "emotional schemata". The latter are different for each individual as they spring from the association of the basic emotions of the sensorimotor level and the situations of individual experience. The schematic level corresponds to spontaneous and true emotions. The conceptual level is based on mechanisms of conscious declarative memory. It stores the abstract notions such as "what are emotions?" or "which situations provoke them?" or "how to deal appropriately with them according to the social rules?" (see Gainotti [Gainotti 2001] for a comprehensive description of hierarchical models).

2.3 Biological models of emotions

The biological models like their cognitive counterparts have either componential or hierarchical organization.

Componential models

Emotions are made up of several components which are subserved by different anatomical structures. With respect to motor expression, the hypothalamus appears to be involved in the generation of autonomic reactions at the more elementary level. Componential models in the neuropsychology of emotions (see clinical implications) prompted two main lines of research. The first considers relationships between disorders that affect different components of the emotional processing system. The second tries to clarify the interactions existing between specific components of emotions and well-defined structures of the brain (see the first case of part two).

Hierarchical models

The biological theories are in line with a phylogenetic perspective. As we saw, hierarchical models organization is based on the complexity of emotional computation performed by different brain structures with control of the highest functional levels over the lowest ones. Brain structures subserving emotions may be based on complexity of operations performed at different levels. The highest brain structures inhibit, modulate and extend rather than replace the lowest and earliest functional systems [Gainotti 2001]. The most influential of these models was proposed by Papez [Papez 1937] who attempted to specify the role played by different brain structures in emotional processing. Emotions were subserved by an anatomical circuit (namely that of Papez) beginning and ending in the hippocampal formation. It included basic structures such as hippocampus, hypothalamus with its mamillary bodies, anterior nuclei of the thalamus, cingulated cortex and their interconnections. This description, which is not completely erroneous, is reconsidered today. New technologies showed that other structures were involved in the Papez's circuit such as the amygdala and the prefrontal cortex. LeDoux and colleagues [LeDoux et al. 1984] showed that the amaygdala (and not the hypothalamus as supposed by Papez) is the structure where in-coming information acquires emotional signification. Another example of a complex model was proposed by Gray and McNaughton [Gray et al. 1996]. They identified specific brain systems: the behavioral inhibition system (BIS), the behavioral approach system (BAS), and the fight-flight system as substrates of these emotional dimensions. According to these authors, the BAS is supposed to regulate approach behavior, sensitivity to reward stimuli and active avoidance behaviour, whereas the BIS is supposed to inhibit instrumental and unconditioned behavior and to control orienting reactions. The authors further hypothesize that structures of different phylogenetic level might mobilize in situations of different complexity the same defensive fight-flight attitude, i.e. when the source of danger is very close and there is no time for analysis or when the situation involves more distant threats and there is more time for analysis. Moreover, the amygdala might mobilize defensive behaviour in light of potential rather than actual events (for a review, see Scherer and Peper [Scherer et al. 2001]). Each of the models presented

above attempt to capture and explain emotion either as a basic or complex process. These models progress consistently over time and become more efficient. For a long time, emotions were considered as useless manifestations, irrational and a source of interference. So, the first cognitivists tried to eliminate emotional dimensions from their models. The former theorists also considered that emotions were undifferentiated. Indeed, it seems that each emotion corresponds to a distinct functional cerebral unit. Two basic emotions, fear and pleasure, were extensively examined over the past years and the studies showed they involve different circuits. The amygdala is involved in the circuit of fear while the accumbens is necessary for pleasure. However, functional neuroimagery studies [Zalla et al. 2000] show that the human amygdala can differentially respond to changes in magnitude of positive or negative reinforcement conveyed by lexical stimuli. Finally, the cognitive models did not take into account the biological constraints provided by the anatomical organization of the brain. Moreover, the adaptive systems subserved by the brain have undergone reorganizations during its phylogenetic history [Tucker et al. 2000]. In the same way, the biological models of emotions reveal some limits:
1) most of the neuroanatomic models are based upon data obtained in animals,
2) divergent neural connectivity subsists across species,
3) furthermore, some controversies exist between the studies in animals and a generalization to humans is sometimes problematic [Scherer et al. 2001].

3 Clinical Implications

Neuropsychology of emotions can be considered to be a very new field of inquiry. The first series of studies conducted in this area were almost exclusively devoted to the problem of hemispheric asymmetries in representation and control of emotions [Borod 1993]. Today, the focus of attention is directed to a much wider array of problems.

3.1 In neuropsychology research (organic diseases)

Alzheimer's disease is a good model for studying the alteration of emotional disturbance because it involves the amygdala which plays an important role in emotions as demonstrated by functional neuroimaging studies [Cahill et al. 1996, Canli et al. 2000]. In the present paper, we shall focus our attention on the relationship between emotions and memory disturbance in AD. In normal controls, emotional items are associated with additional semantic or autobiographical indices when they learn new information. Emotions may serve as a retrieval cue. A person may initially remember how they felt about an event, and that cue may then allow them to generate additional features about the event. Alzheimer's disease (AD) results in significant atrophy of the medial temporal lobe that leads to a dramatic memory deficit. At the early stage, impairment concerns mainly memory which is characterized by an inability to learn new information or to recall previously learned information. Memory decline is gradual and the disease produces other cognitive deficits as it progresses. Emotional influence on recall was studied by means of cognitive neuropsychology of memory [Tulving 1972]. This model postulates that memory can be considered in terms of dissociable systems, distinct processes, and neuroanatomical structures.

Within long-term memory systems, episodic memory (i.e. knowledge of episodes and facts that can be consciously recalled and related) is typically severely impaired in early-stage of the disease. Semantic memory that underlies knowledge and language is less likely to be significantly affected, although impairment may be observed in some individuals. Procedural memory (i.e. ability to gradually acquire and retain motor, perceptual and cognitive skills) is preserved, as are some aspects of priming. Memory can also be considered in terms of the processes of encoding, storage and retrieval [Tulving et al. 1973].

Some studies [Kazui et al. 2000, Moayeri et al. 2000] showed that recall of AD patients is typically better for emotional than for neutral stimuli. Memory is also better for neutral stimuli embedded in an emotional context. Other studies, in contrast, concluded that AD disrupts memory enhancement for verbal emotional information [Hamann el al. 2000, Kensinger et al. 2004]. AD patients also demonstrated impairments in emotionally mediated implicit memory (Padovan et al. 2002). Differences across studies are related to the heterogeneity of patient populations (disease severity), difference in the stimuli or the extent of amygdala atrophy [Mori et al. 1999]. Emotional arousal improves episodic memory in patients with AD and gives a clue to the management of people with dementia [Kazui et al. 2000]. Rehabilitation of emotions is based on aspects of emotional communication such as prosody. A series of experiments [Thaut et al. 2005] investigated the effect of music as a mnemonic device on learning and memory. More researches are needed to develop a useful strategy for memory improvement.

3.2 In cognitive and behaviour therapy (functional disorders)

Panic disorder is relevant for studying fear, a basic emotion. It can be depicted as a profound blast of anxious affect. The physical symptoms are multiple: shortness of breath, rapid heart rate, dizziness, tingling, sweating. The cognitive symptoms involve automatic thoughts and mental images which tend to be catastrophic, i.e. there is a tendency to exaggerate the dangerousness of a situation and simultaneously to underestimate the control over the danger. Therapeutic actions are based on classical conditioning theories [Pavlov 1928]. Fear conditioning occurs when initially innocuous conditioned stimulus (CS) is associated with an aversive unconditioned stimulus (US) that activates unconditioned fear responses (URs). The CS comes to elicit various conditioned responses (CRs) that share similar characteristics to innate fear responses [Kim et al. 2006]. The best known example of fear conditioning reported by the authors is the little Albert case [Watson et al. 1920]. As the authors recount: "Albert was an infant who initially exhibited curiosity over a white rat by touching and playing with it. As Albert's hand touched the rat, the experimenters triggered a big noise behind his head (US) causing him to startle and cry (UR). Afterwards, when the rat (CS) was placed near Albert's hand, he withdrew his hand and began to cry (CR). This exhibition of fear towards the rat was generalized to other white furry animals and objects."
Treatment of patients suffering from panic disorder involves exposure to fear cues (behaviour therapy) and cognitive restructuring (cognitive therapy). One powerful means of reducing anxiety problems is believed that of countering avoidance. Avoidance reduces anxiety in short term, but makes for more anxiety in the long term as

avoidance increases over time. *Exposure* involves placing someone in the avoided situation until the anxiety decreases completely. The disappearance of anxiety is called *Extinction*. *Cognitive restructuring* is used to identify and counter fear of bodily sensations. Patients are encouraged to consider the evidence and think of alternative possible outcomes following the experience of bodily cues.
LeDoux's model [LeDoux 1986] provides a theoretical framework for therapeutic actions of cognitive-behavioural therapy as it establishes a relationship between emotions and cognitive factors. The model postulates the contribution of the thalamus and the amygdala in fear conditioning and anxious reactions. These structures form a circuit that involves immediate survival responses (i.e. flight or fight reactions). The connection between the thalamus and the amygdala is the most direct and therefore the fastest. In parallel, another pathway exists which includes the prefrontal cortex and the hippocampus in addition to the thalamus and the amygdala. The prefrontal cortex is involved in a more complex conditioning requiring the planning of actions. An individual takes more time for cognition to shift from reaction to action and he is seen as an emotional actor who copes with a cognitive plan of voluntary action rather than just a reactor to an involuntarily elicited emotional reaction. In the light of this model, it is easier to understand the therapeutic actions of exposure to fear cues and cognitive restructuring. Exposure to fear cues seems to be mediated by the more direct circuit while cognitive restructuring, which explores several option-responses seems to be supported by the second longer pathway, which involves the prefrontal cortex. Extinction of anxiety is explained by the action of prefrontal cortex. Bodily feedback of panic disorder (i.e. shortness of breath, rapid heart rate, sweating) is also taken into account in this model. Somatic responses are stored in amygdala together with their associated perceptual context. When patients experience bodily feedback of fear, they misinterpret them and tend to develop avoidance behaviour.

4 Conclusion

In conclusion, theories produce important results in the field of cognitive neuropsychology and provide a better understanding of brain functioning. This comprehension leads to profound changes in clinical practice in the evaluation of patients and produces new orientations in the rehabilitation. Neuropsychologists progressively changed their conception of the brain that is more theoretically and methodologically constructed. Single case studies have been crucial in the historical development of the discipline and they continue to lend consistent support to the development of existing models or in the elaboration of new ones. Models of cognitive and behaviour therapy give a theoretical framework for therapeutic actions and lead to new perspectives of treatment of patients. The exchange between theory and clinical practice allows progressive adjustment between the knowledge of the brain functioning and the management of patients.

References

[Borod 1993] Borod, J. (1993) *Cerebral mechanisms underlying facial, prosodic, and lexical emotional expression: a review of neuropsychological studies and methodological issues.* Neuropsychology, 12, 2493-2503.

[Cahill et al. 1996] Cahill, L.; R.J. Haier; J. Fallon; M.T. Alkire; C. Tang; D. Keator; J. Wu; and J.L. McGaugh. (1996) *Amygdala activity at encoding correlated with long-term, free recall of emotional information.* Processings of the National Academy of Sciences USA, 93, 8016-21.

[Canli et al. 2000] Canli, T.; Z. Zhao; J. Brewer; J.D. Gabrieli; and L. Cahill. (2000) *Event-related activation in the human amygdala associates with later memory for individual emotional experience.* Journal of Neuroscience, 20, 1-5.

[Ekman 1984] Ekman, P. (1984) *Expression and the nature of emotion.* In Approaches to Emotion, K.R. Scherer and P. Ekman (Eds.). Hillsdale, NJ: Erlbaum, 319-344.

[Gainotti 2001] Gainotti, G. (2001) *Emotions as a biologically adaptive system: an introduction,* In Emotional behaviour and its disorders, F. Boller and J. Grafman (Eds.). Handbook of Neuropsychology, Elsevier, Amsterdam, 1-15.

[Gray et al. 1996] Gray J.A. and N. McNaughton (1996) *The neuropsychology of anxiety: reprise.* In Nebraska Symposium on Motivation: Perspectives on Anxiety, Panic and Fear, D.A. Hope (Ed.). Lincoln, NE: University of Nebraska Press, 61-134.

[Hamann el al. 2000] Hamann, S.B.; E.S. Monarch; and F.C. Goldstein (2000) *Memory enhancement for emotional stimuli is impaired in early Alzheimer's disease.* Neuropsychology, 14, 82-92

[Kazui et al. 2000] Kazui, H; E. Mori; M. Hashimoto; N. Hirono; T. Imamura; S. Tanimukai; T. Hanihara; and L. Cahill (2000) *Impact of emotion on memory. Controlled study of the influence of emotionally charged material on declarative memory in Alzheimer's disease.* British Journal of Psychiatry, 177, 343-7.

[Kensinger et al. 2004] Kensinger, E.A.; B. Brierley; N. Medford; J.H. Growdon; and S. Corkin. 2004. *Effects of Alzheimer disease on memory for verbal emotional information.* Neuropsychologia, 42, 791-800.

[Kim et al. 2006] Kim, J.J.; and M.W. Jung (2006) *Neural circuits and mechanisms involved in Pavlovian fear conditioning: a critical review.* Neuroscience and Biobehavioral Reviews, 30, 188-202.

[Lazarus 1982] Lazarus, R.S. (1982) *Thoughts on relations between emotion and cognition.* American Psychologist, 37, 1014-1019.

[LeDoux et al. 1984] LeDoux, J.E.; A. Sakagachi; and D.J. Reis (1984) *Subcortical efferent projections of the medial geniculate nucleus mediate emotional response conditioned by acoustic stimuli.* Journal of Neuroscience 4, 683-689.

[LeDoux 1986] LeDoux, J.E. (1986) *Cognitive & emotional interactions in the brain.* Cognition and Emotion, 3, 267-289.

[Moayeri et al. 2000] Moayeri, S.E.; L. Cahill. Y. Jin; and S.G. Potkin (2000) *Relative sparing of emotionally influenced memory in Alzheimer's disease.* Neuroreport, 11, 653-5.

[Mori et al. 1999] Mori, E.; M. Ikeda; N. Hirono; H. Kitagaki; T. Imamura; and T. Shimomura (1999) *Amygdalar volume and emotional memory in Alzheimer's disease.* 156, 216-22.

[Padovan et al. 2002] Padovan, C.; R. Versace; C. Thomas-Anterion; and B. Laurent (2002) *Evidence for a selective deficit in automatic activation of positive information in patients with Alzheimer's disease in an affective priming paradigm.* Neuropsychologia, 40, 335-9.

[Papez 1937] Papez, J.W. (1937) *A proposed mechanism of emotion.* Archives of Neurology and Psychiatry, 79, 217-224.

[Pavlov 1928] Pavlov, I.P. (1928) "Lectures on conditoned reflexes" New York International.

[Scherer 1984] Scherer, K.R. (1984) *On the nature and function of emotion. A component process.* In Approaches to Emotion, K.R. Scherer and P. Ekman (Eds.). Hillsdale, NJ: Erlbaum, 293-318.

[Scherer et al. 2001] Scherer, K.R.; and M Peper (2001) *Pscyhological theories of emotion and neuropsychology research* In Emotional behaviour and its disorders, F. Boller and J. Grafman (Eds.). Handbook of Neuropsychology, Elsevier, Amsterdam, 17-48.

[Schorr 2001] Schorr, A. (2001) *Appraisal : the evolution of an idea. In Appraisal Processes in Emotion: Theory, Methods, Reseach.* K.R. Scherer, A. Schorr and T. Johnstone (Eds.). Oxford: Oxford Univeristy Press, 20-34.

[Thaut et al. 2005] Thaut, M.H.; D.A. Peterson; and G.C. McIntosh (2005) *Temporal entrainment of cognitive functions: musical mnemonics induce brain plasticity and oscillatory synchrony in neural networks underlying memory* Annals of the New York Academy of Sciences, 1060, 243-54

[Tucker et al. 2000] Tucker, D.M.; D. Derryberry; and P. Luu (2000) *Anatomy and physiology of human emotion: vertical integration of brainstem, limbic and cortical systems.* In The neuropsychology of emotion, J.C. Borod (Ed.). New York, Oxford University Press, 56-79.

[Tulving 1972] Tulving, E. (1972) *Episodic and semantic memory.* In: Organisation of memory, E. Tulving and W. Donaldson (Eds.). Academic Press, New York, 381-403.

[Tulving et al. 1973] Tulving, E.; and D.M. Thomson (1973) *Encoding specificity and retrieval processes in episodic memory.* Psychological Review, 80, 352-373.

[Watson et al. 1920] Watson, J.B. and R.R. Rayner (1920) *Conditioned emotional reaction* Journal of Experimental Psychology, 3, 1-14.

[Zajonc 1980] Zajonc, R.B. (1980) *Feeling and thinking: preferences need no inferences.* American Psychologist, 2, 151-176.

[Zalla et al. 2000] Zalla, T.; E. Koechlin; P. Pietrini; G. Basso; P. Aquino; A. Sirigu; and J. Grafman (2000) *Differential amygdala responses to winning and losing: a functional magnetic resonance imaging study in humans* European Journal of Neuroscience, 12, 1764-70.

Part V

Production systems and simulation

Complex Systems Dynamics in an Economic Model with Mean Field Interactions

Gianfranco Giulioni

Dipartimento di Metodi Quantitativi e Teoria Economica,
Viale Pindaro 42,
65127 Pescara, Italy
g.giulioni@unich.it

Key words: Economy as a complex systems, macroeconomic dynamics, complex dynamics, attractors.

Summary. In this paper we built and simulate a model where the economy is viewed as a complex system. Our economy is composed of a high number of different agents that are interacting in an indirect (mean field) way, but not through a coordination device. The resulting macroeconomic dynamics can be defined as complex because they change attractor in time without no relevant change in the parameter values (we can also say that the degree of self-organization observed in our system changes with time). One of these attractors is a limit cycle and consequently our work can be viewed as a contribution to the endogenous business cycle theory.

1 Introduction

One of the most important results of neoclassical Economics, the General Equilibrium Theory, relies on the existence of a coordination mechanism introduced using the elegant device of the walrasian auctioneer. This is probably a provoking sentence, but it opens an important debate in the economic profession: is the presence of coordination mechanisms a good approximation of the economic reality? A positive answer to this question would prevent studies in economics to enter roads already opened for other disciplines like those of the self-organization phenomena and complexity theory. Fortunately, in recent years, a small but growing number of economists became convinced that the economy is a complex system (see [Anderson et al. 1988, Arthur et al. 1997, Blume et al. 2005] for example) and therefore started to travel these roads.

Before preparing for the trip proposed in this paper (of course traveling the road we are talking about) some preliminary comments and definitions are useful. In our view, a complex system is composed of a high number of different elements that are interacting in some way, but not through a coordination device. Complex systems are interesting because under certain conditions they give rise to "unusual" dynamics. In particular it is possible that while one of the parameters changes smoothly, the behavior of some endogenous macroscopic variable changes in an unexpectedly organized way or, putting it another way, structures form in an unstructured environment. When such structures emerge without a coordination device researchers generally say that the system "self-organizes". Sometimes, the expression self-organization is used to denote the emergence of structures in the phase space of dynamical systems. Indeed systems composed of a low number of difference or differential equations can display more structured attractors on the phase space when a parameter is gradually moved. This is not the way the expression self-organization is used in this paper. Dynamical systems are intractable when their dimension increases and consequently they cannot be classified as complex systems (recall that according to our definition a complex systems has a very high dimension). Talking about dynamical systems a confusion may arise because a dynamical system (that, from our point of view, is not complex) can display chaotic dynamics that are usually referred to as "complex dynamics". Thus, despite the similarity of the expressions, in what follows, "complex systems dynamics" has a different meaning from "complex dynamics". In particular the latter are a subset of the former at least as long as one identifies chaotic dynamics with the complex dynamics. More interestingly complex systems may exhibit dynamics never detected in dynamical systems. In cellular automata systems, for instance, the existence of such a type of dynamics was found by [Wolfram 1986], [Langton 1986], and [Packard 1988]. The last one coined the expression "the edge of chaos" to identify them. We will refer to this type of dynamics as "complex systems dynamics".

The aim of this paper is to show how the economic system can give rise to "complex systems dynamics." The model presented below belongs to a set born out of a paper by [Greenwald et al. 1993] (GS hereafter). The intent of these works is to show how the financial conditions of firms is a determinant of the aggregate production of countries, that is, of the Gross Domestic Product (GDP). It is well known that GDP has cyclical dynamics (indeed the explanation of this phenomenon is one of the main topics of macroeconomic theory) and, from a dynamical systems point of view, this calls for the presence of a limit cycle in some relevant variable. GS obtain a difference equation for the financial condition of firms (represented by the equity base) that, under certain parameterization, gives rise to limit cycles and to chaotic dynamics. From our point of view, GS's work has the inconveniece that the millions heterogeneous firms populating the economy are replaced by one of them that is supposed to be representative.[1] This way to proceed is questionable because, among other drawbacks (see [Kirman 1992] for example), it limits the analysis to the use of dynamical systems tools that, as maintained above, shuts out complex systems dynamics. Dynamical systems theory is also used in a paper by [Delli Gatti et al. 2000]. Building on GS they go a step further stressing the importance of heterogeneity, but they take it into account introducing a difference equation for the variance of the fi-

[1] GS are aware of the importance of heterogeneity. In the cited paper they use the representative agent because the focus is on showing how financial factors affect the aggregate production rather than on analizing the effects of heterogeneity.

nancial position ending up with the analysis of a two dimensional dynamical system.

In more recent times, GS's type of model have been analyzed using agent based simulation techniques that is, according to us, a more convenient way to deal with complex systems. There are no equations for the macroscopic variables, but only equations governing the individuals' behavior. The values of the macroscopic variables are recovered by simply summing or averaging the individuals' ones. Consequently it may happen that the dynamics at the macroscopic level are completely different from those at the individual level identifying genuine emergent phenomena. [Delli Gatti et al. 2005] for example show how one can recover particular statistical distributions (basically they are fat tailed distributions like power laws or Weibull) out of the individual data or from the aggregate time series obtained from simulations, and that the same distributions characterize real data. The important observation is that according to a number of scientists the presence of these distributions is common in complex systems dynamics (see [Bak 1997] for example).

In what follows we build a model using some "ingredients" from the above cited papers. We then report some simulation results showing how the model produces peculiar dynamics that could be defined as "complex systems dynamics".

2 The model

The economy is populated by a large number of firms. As in the GS's type of model we concentrate our attention on the firm. Consumers and others economic agents are supposed to passively accommodate firms' decisions. In these supply side models the production function has a very important role (a large part of the macroeconomic theory is of the supply side type, think for example of the exogenous and endogenous growth and of the real business cycle theories). In the present model, the production function is linear:

$$Y_{it} = \nu_{it} K_{it}$$

where Y_{it} is the production, K_{it} the capital and ν_{it} its productivity.

We dedicate the remainder of this section to the two determinants of the production: ν and K. As mentioned before, we are interested in the emergent properties of the aggregate production dynamics $Y_t = \sum_i Y_{it}$ that we'll recover using a bottom-up approach, that is, through agent based simulations.

Some preliminary notions on the firms variables will be useful and are given here. The balance sheet of a firm is $K_{it} = D_{it} + A_{it}$. where D_{it} is debt and A_{it} the equity base. The fraction $\frac{A_{it}}{K_{it}} = a_{it}$ is the equity ratio that is a signal of the financial soundness of the firm. The dynamics of the balance sheet variables are strictly relative to the economic result of the firm (π_{it}). In these preliminary notions, we restrict ourselves to note that this variable directly affects the dynamics of the equity base in the following way: $A_{it} = A_{it-1} + (1 - \eta_{it})\pi_{it}$ where η_{it} is the fraction of the economic result that does not affect the equity base (more detailed explanation below). The important aspect is that the economic result can be negative and this decreases the equity base. As a consequence, the equity base of a firm could become negative and, if this happens, the firm must leave the market. Another exit mechanism will be considered and we'll come back to this issue below, but what is important to note here is that the presence of exit mechanisms calls for the existence of an entry process. These considerations serve to highlight that an important aspect of this kind of model is the firms turnover [Delli Gatti et al. 2003]. However, in this paper we avoid such complication adopting the one-to-one replacement assumption (each exiting firm is replaced by a new one).

Now we can look at the model description starting from the economic result of the firm. In the following steps we use the economic result to determine the dynamics of the two variables we are interested in: the capital (K_{it}) and the productivity (ν_{it}).

2.1 Economic result

The economic result (π_{it}) is given by revenues (R_{it}) minus costs (C_{it}):

$$\pi_{it} = R_{it} - C_{it} \tag{1}$$

All the variables are in real terms so that prices will never appear in our equations.

Revenues.

Firms sell all their product, but their real revenue from sales may be different from the production due to unforeseen external events (in GS for example this is due to an unknown selling price). We formulate this as follows

$$R_{it} = \nu_{it} K_{it} + u_{it} K_{it} \tag{2}$$

where u is a random variable with mean equal to 0 and finite variance.

Costs.

Costs are of two types: production costs (C^L) and adjustment costs (C^K)

$$C_{it} = C_{it}^L + C_{it}^K \tag{3}$$

Production costs. Production costs are due to labor. We use the simplifying assumption that firms need one worker for each unit of capital (Leontief type production function) so that $L_{it} = K_{it}$. Labor costs are

$$C_{it}^L = w_{it} L_{it} = w_{it} K_{it} \tag{4}$$

where w_{it} is the wage and L_{it} the number of employed workers.

Adjustment costs. The adjustment costs must be sustained to adapt the stock of capital [Mussa 1977]. We use here the formulation adopted in [Delli Gatti et al. 2000]

$$C_{it}^K = \frac{\gamma}{2} \frac{(K_{it} - K_{it-1})^2}{\bar{K}_{t-1}} \tag{5}$$

where $\bar{K}$ is the average capital of the economy. This introduces a first mean field interaction in the model.

The economic result.

using equations (1)-(5) the economic result is

$$\pi_{it} = \nu_{it} K_{it} - w_{it} K_{it} - \frac{\gamma}{2} \frac{(K_{it} - K_{it-1})^2}{\bar{K}_{t-1}} + u_{it} K_{it}$$

note that because of the Leontievian assumption, ν_{it} is also the labor productivity. We assume that the wage is related to the (latest known) average labor productivity. From this base we use the following assumption: $w_{it} = \bar{\nu}_{t-1}$ that introduces a second mean field interaction being $\bar{\nu}_{t-1}$ the average productivity of the period before. Under this assumption we can write

$$\pi_{it} = (\nu_{it} - \bar{\nu}_{t-1}) K_{it} - \frac{\gamma}{2} \frac{(K_{it} - K_{it-1})^2}{\bar{K}_{t-1}} + u_{it} K_{it} \tag{6}$$

In order to simplify, we avoid discussing the effects of changing the capital level on the economic result. A discussion of these aspects would reveal that the investment involves the movement of the debt stock and affect minimally the economic result. This effect does not modify the behavior of the system and can be eliminated under a further simplifying assumption.

2.2 The evolution of Capital

To choose the optimal level of capital the firm maximizes the economic result function, but with two changes. First of all, at the time of the choice, firms don't know the realization of the random variable so that it is replaced with the average value. This allows us to omit the term $u_{it} K_{it}$ being the mean of u equal to zero. Secondly we assume that firms don't know also the average capital and replace it with their own level of capital. Consequently, the objective function to maximize is

$$E(\pi_{it}) = (\nu_{it} - \bar{\nu}_{t-1}) K_{it} - \frac{\gamma}{2} \frac{(K_{it} - K_{it-1})^2}{K_{it-1}}$$

Maximizing with respect to K_{it} we have the first element we need, that is, the dynamics of the capital

$$K_{it} = \frac{\nu_{it} - \bar{\nu}_{t-1}}{\gamma} K_{it-1} + K_{it-1} \tag{7}$$

2.3 The dynamics of the productivity

The second element we need is the dynamics of the productivity ν_{it}. This variable moves if the firm funds Research and Development activities.

Investment in Research and Development activities.

At the end of the period the economic result is realized. It can be positive (profit) or negative (loss).

When a profit is realized the firm has to decide how to use it. We assume that it can be used other than to increase the equity base, to finance Research and Development (R&D) activities. In particular R&D investments are assumed to be

$$R\&D_{it} = \pi_{it}\eta_{it}$$

where η_{it} is the share of profit dedicated to R&D. We assume that this share is an increasing function of the financial soundness of the firm represented by the equity ratio as follows:

$$\eta_{it} = \begin{cases} a_{it-1} & \text{if } \pi_{it} > 0 \\ 0 & \text{if } \pi_{it} \leq 0 \end{cases}$$

From these considerations we can also recover the dynamics of the equity base:

$$A_{it} = A_{it-1} + (1 - \eta_{it})\pi_{it} \tag{8}$$

The dynamics of the productivity.

The outcome of the R&D investment is stochastic and the probability of success increases with the amount of funds dedicated to these activities. We formulate this probability as

$$pr = \frac{1}{1 + e^{-b(R\&D_{it} - c)}}$$

where b and c are parameters.

If a firm obtains a success from its R&D activities, its productivity increases by a constant amount β, so that the dynamic of the productivity is

$$\begin{cases} \nu_{it+1} = \nu_{it} + \beta & \text{with probability } pr \\ \nu_{it+1} = \nu_{it} & \text{with probability } 1 - pr \end{cases} \tag{9}$$

3 Simulations

We simulate the model using object oriented programming languages. In a first implementation the objective-C version of the SWARM library is used. The validity of the results is checked coding a second time the same model using the RePast java library. We run a large number of experiments to check how the model reacts to changes in the initial conditions, the size of the system (that is the number of firms) and the parameters. Among them, the parameter γ has a very important role. For high values of this parameter the system displays a limit cycle, while cycles disappear for low values. In between, there is a non negligible region where the system gives rise to complex systems dynamics.

We describe here in details one of these experiments where we set $\gamma = 1.5$. The comments below serve also to better explain how the model works. At the beginning the code creates 100000 identical firms giving them the following initial conditions: $K_{i0} = 100$, $A_{i0} = 30$, $\nu_{i0} = 0.1$. The parameters b, c and β are set to 3, 2 and 0.01, respectively.

The algorithm goes through the following steps:

```
1 reset values to firms that meet the exit conditions
2 update the capital
3 update equity ratio
4 update the profit using the random variable
5 update investments in R&D
6 update productivity
7 update equity base
8 collect data
```

Because some steps are technical, we discuss the flow of events in a logical order, this implies that the order reported above will not be respected on some occasions. First of all firms decide their new capital level (step 2) using equation (7): firms with a productivity higher than the average increase capital while the others reduce it. Firms employ the new stock of capital in the production and realize a production equal to $\nu_{it}K_{it}$. Once the production is realized it is sold on the market. The average revenues from sales is equal to production, but some firms realize a higher revenue and some others a lower one due to contingent situations represented by the random variable u that is supposed to be uniform with bounds -0.1 and 0.1. Now, with revenues in their hands the entrepreneurs have to pay their costs: wages and adjustment costs. Here two situations are possible. In the first one, the revenue from sales is higher than costs and consequently a profit is realized. In the second, the revenue from sales is lower than costs and the firm suffers a loss. This is the content of equation (6) implemented in step 4. In step 5 firms with a profit spend a share equal to their equity ratio (that is calculated in step 3) of profit in R&D, while firms that suffer a loss make no expenditures in R&D. After this computation we know for each firm how much they spend in R&D. This allows us to update the productivity using equation (9) in step 6. Moreover, knowing R&D expenditures allows us to determine how the economic result affects the equity base. We do this coding equation (8) in step 7. Finally we record data (step 8) and a new iteration is about to start. At the beginning of the new iteration (step 1), we check the value of the equity base computed in step 7. If it is negative the variables of the firm are initialized with the following values $K_{it} = K_{i0} = 100$, $A_{it} = A_{i0} = 30$, $\nu_{it} = \bar{\nu}_{t-1}$. This can happen to firms that suffer a loss in step 4 of the previous iteration. The fact that they suffer a loss means that they are not able to cover costs with the revenues from sales. At this point they must resort to their internal funds represented by the equity base. In some cases even the equity base is not sufficient to provide the additional needed funds and the firm must exit the market. In addition to this exit mechanism we add also a threshold to the size of the firm, that is, firms with a low level of capital ($K_{it-1} < 20$) but with a positive equity base are also replaced. However, this second exit mechanism is present to catch exceptions and does not affect the simulation results. Finally, note that resetting the variables of the firms when they meet these conditions is the same as assumuming a one-to-one replacement situation and the number of firms is constant to 100000 during the whole simulation.

The results are showed in the following graphs and commented on in next section.

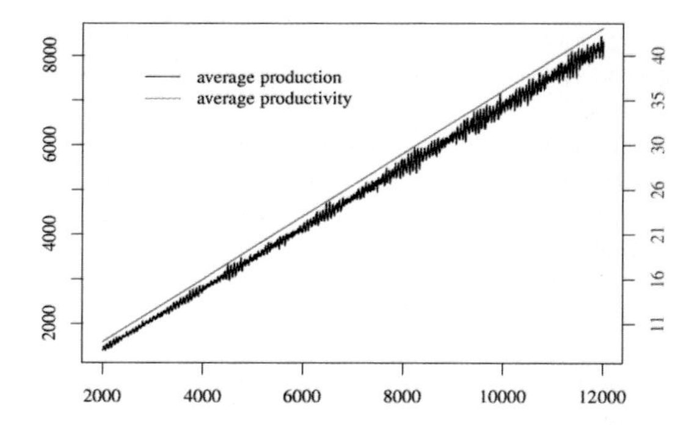

Fig. 1. Average production (left axis and black line) and average productivity (right axis and gray line). 2000 time steps have been discarded to eliminate the transient state

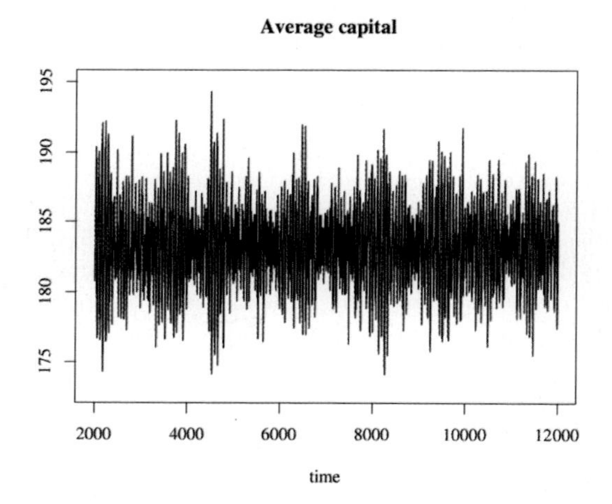

Fig. 2. Average Capital. 2000 time steps have been discarded to eliminate the transient state

4 Discussion

Although the reported graphs could contain interesting features from the economic point of view, the focus of the discussion will be mainly on the type of dynamics a system like this can generate.

Figures 1 and 2 show the dynamics of the variables involved in the production function (production, capital and productivity) for 10000 simulation time steps. We don't show the initial 2000 time steps to avoid the transient state (it occurs in decrising oscillations). In the graphs the average values are reported. Regarding production and capital, one might be interested in the aggregate values. They can be obtained by multiplying those reported in the graphs for the number of firms in the economy that in our case is fixed and equal to 100000. Consequently, the qualitative behaviors of the aggregate production and capital are exactly the same as those reported in the graphs. First of all, from figure 1 it is evident that the increasing trend in production is due to the increasing productivity. Secondly, it is also evident that production is much more volatile than productivity. Having the production function in mind it is straightforward that the volatility of production depends on that of capital; figure 2 confirms this deduction. It is also easy to see that the volatility is not constant, but changes with time in an irregular way. This pushes us toward a more accurate investigation. Figures 1 and 2 report too much data to get an insight into the nature of the volatility by visual inspection.

Figures 3, 4 and 5 shed some light on the phenomenon. In these graphs three contiguous sub-periods with different volatilities are shown. Figure 3 shows that, in the time span 7750-8750, we are not dealing with stochastic volatility but with a more structured behavior: limit cycles. This is surprising because it is the result of an agent based model with idiosyncratic shocks. As discussed above, obtaining a limit cycle in a dynamical system in not hard, but dynamical systems contemplate the presence of a very low number of equations. In the present model the cyclical behavior is obtained averaging a large number of stochastic equations (one for each firm). From a probability theory point of view, what is reported in the figure is an average of a large number of identical stochastic processes. Figure 3 suggests that the law of large numbers, according to which one expects a very smooth behavior of the average value, does not hold at list in the reported periods. Furthermore, we cannot maintain that this is a feature of the individual behavior preserved at the aggregate level. The uncorrelated idiosyncratic shock present in the model differentiates firms' decisions and, from this point of view, the law of large numbers should apply. A cyclical behavior of the average requires that the various components of the system act in a strong correlated way that, in the absence of a representative agent, could be possible if a coordination mechanism were contemplated. But here we have no coordination device, here each entrepreneur decides alone using its private information (capital and productivity) and the average level of the productivity. Our final conjecture is that in an agent based model the presence of a replacement process and that of mean field interactions [Aoki 1996] can give rise to a considerable degree of self-organization.

A second observation comes from comparing Figures 3, 4, and 5: the system seems to be able to change attractor in time. The dynamic presented in Figure 4 is quite different from the ones visible in Figures 3 and 5 although they were obtained with the same parameters. In Figure 4 the law of large numbers seems to have a stronger effect than in the other two graphs, that is, the degree of self-organization changes with time. At the actual state of the investigation this phenomenon seems to be a deep emergent property of the system. Indeed one candidate for the explanation could be the average productivity (because it is not fluctuating around a constant value), but looking at its smooth behavior, it seems hard to give it the responsibility to change the system behavior from a limit cycle to (something similar to) an equilibrium and then back to a limit cycle.

The changes in the attractor are also showed in Figure 6 where average values for very short time spans (they are sub-periods of Figure 3 and 4) are showed in the equity ratio-capital phase space. It is evident how the economic dynamics can

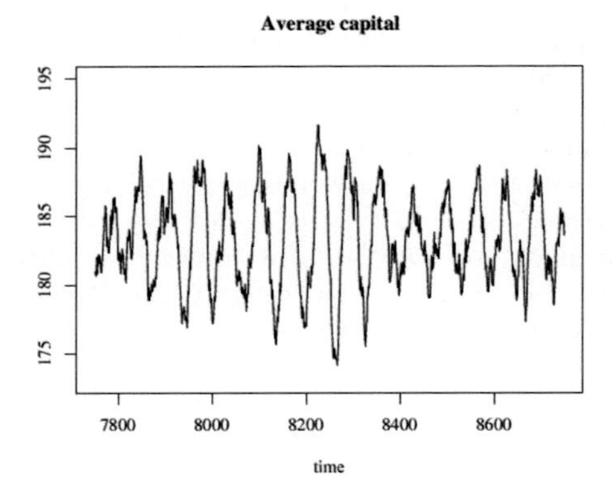

Fig. 3. Average Capital from time 7750 to 8750

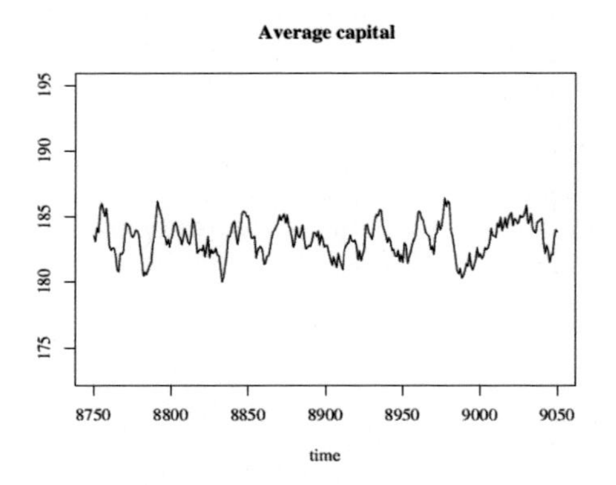

Fig. 4. Average Capital from time 8751 to 9050

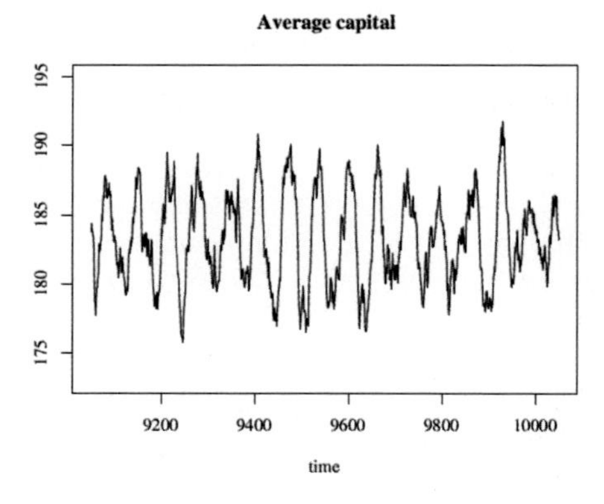

Fig. 5. Average Capital from time 9051 to 10050

commute between simple (as in the time span 8850-8920) to more structured (as in the time span 8220-8290) attractors. This Figure is also interesting from the economic point of view. Indeed, as discussed in the introduction, GS's type models prove that there is a relationship between the aggregate production and the financial soundness of the economy. Indeed the Figure shows that this relationship exists and is strong in some time spans. Furthermore, looking at the black line in Figure 6, it could be maintained that production and financial fragility move as described by [Minsky 1982] in his financial fragility theory of macroeconomic fluctuations. On the other hand, this behavior is not always so strong to be detected as the gray line of the Figure shows.

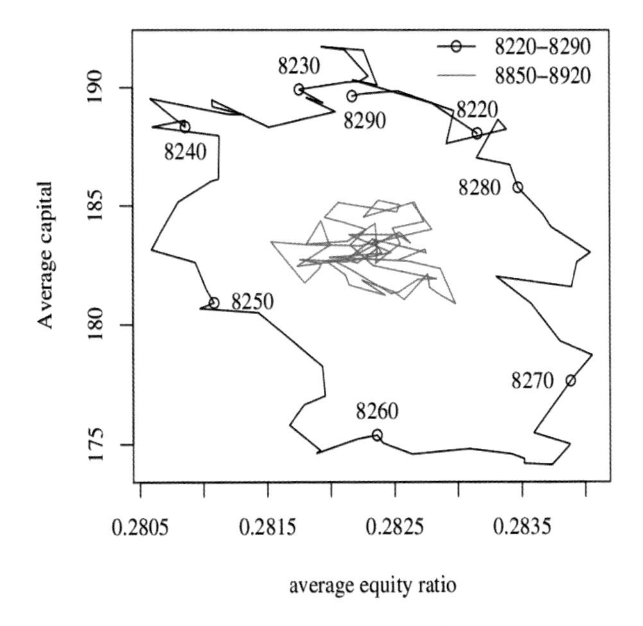

Fig. 6. Different attractors in two different time spans

References

[Anderson et al. 1988] Anderson, P. W.; K. J. Arrow, and D. Pines, editors (1988) *The Economy as an Evolving Complex System.* New York,Addison-Wesley.

[Aoki 1996] Aoki, M. *New Aprroaches to Macroeconomic Modeling.* Cambridge University Press, Cambridge, 1996.

[Arthur et al. 1997] Arthur, B. Z.; S. N. Durlauf, and D. W. Lane, editors (1997) *The Economy as an Evolving Complex System II.* Addison-Wesley.

[Bak 1997] Bak, P. (1997) *How Nature Works. The science of Self-Organized Criticality.* Oxford University Press, Oxford.

[Blume et al. 2005] Blume, L. E. and S. N. Durlauf, editors (2005) *The Economy As an Evolving Complex System, III : Current Perspectives and Future Directions.* Addison-Wesley.

[Delli Gatti et al. 2005] Delli Gatti, D.; C. Diguilmi, E. Gaffeo, M. Gallegati, G. Giulioni, and A. Palestrini (2005) *A new approach to business fluctuations: heterogeneous interacting agents, scaling laws and financial fragility.* Journal of Economic Behavior and Organization, 56, 489–512.

[Delli Gatti et al. 2003] Delli Gatti, D.; M. Gallegati, G. Giulioni, and A. Palestrini (2003) *Financial Fragility, Patterns of Firms' Entry and Exit and Aggregate Dynamics.* Journal of Economic Behavior and Organization, 51, 79–97.

[Delli Gatti et al. 2000] Delli Gatti, D.; M. Gallegati, and A. Palestrini (2000) *Agent's Heterogeneity, Aggregation and Economic Fluctuation.* In D. Delli Gatti, M. Gallegati, and A. P. Kirman, editors, *Interaction and Market Structure*, Berlin,Springer.

[Greenwald et al. 1993] Greenwald, B. C. and J. E. Stiglitz (1993) *Financial Market imperfections and Business Cycles.* Quarterly Journal of Economics, 108, 77–114.

[Kirman 1992] Kirman, A. P. (1992) *Whom or What Does The Representative Individual Represent.* Journal of Economic Perspective, 6:117–36.

[Langton 1986] Langton, C. (1986) *Studying Artificial Life with Cellular Automata.* Physica D, 2, 120–149.

[Minsky 1982] Minsky, H. P. (1982) *The financial instability hypothesis: Capitalist processes and the behavior of the economy.* In C. P. Kindleberger and J. P. Laffargue, editors, *Financial Crises: Theory, History and Policy.* Cambridge University Press, Cambridge.

[Mussa 1977] Mussa, M. L. (1977) *External and Internal Adjustment Costs and the Theory of Aggregate and Firm Investment.* Economica, 44, 163–178.

[Packard 1988] Packard, N. (1988) *Adaptation Toward the Edge of Chaos.* Technical report, Center for Complex Systems Research, University of Illinois.

[Wolfram 1986] Wolfram, S. (1986) *Theory and Applications of Cellular Automata.* World Scientific, Singapore.

Complexity of Traffic Interactions: Improving Behavioural Intelligence in Driving Simulation Scenarios

Abs Dumbuya[1], Anna Booth[1], Nick Reed[1], Andrew Kirkham[1], Toby Philpott[1], John Zhao[2], and Robert Wood[2]

[1] TRL, Crowthorne House
Nine Mile Ride, Wokingham
Berkshire, RG40 3GA, UK
adumbuya@trl.co.uk

[2] Loughborough University
Loughborough, Leicestershire
LE11 3TU, UK

Summary. This paper introduces modelling concepts and techniques for improving behavioural intelligence and realism in driving simulation scenarios. Neural Driver Agents were developed to learn and successfully replicate human lane changing behaviour based on data collected from the TRL car simulator.

Key words: AI-supported simulation, neural network, simulators, Behavioural science, Psychology

1 Introduction

The design and development of realistic scenarios for driving simulators (see discussion on emerging issues relating to the realism of visual databases and scenarios, [Parkes 2005, Allen et al. 2003, Allen at al. 2004]) could greatly enhance the realism with which simulator trials can be created, since the autonomous vehicles would be capable of responding in a realistic manner both to the behavioural responses of the participant and to any pre-programmed autonomous vehicle behaviour (e.g. a vehicle programmed to disobey a red traffic light). This in turn would improve participants' immersion in simulator scenarios, increasing the likelihood that they will drive in a realistic and representative manner with the consequence that greater confidence can be placed in resulting analyses. This paper describes a research project at TRL which extended previous work (development of a Synthetic Driving SIMulation, SD-SIM framework) conducted at Loughborough University [Dumbuya et al.

2002, Dumbuya et al. 2003]. The paper demonstrates the development and application of a novel technique for improving and verifying the realism of a Neural Driver Agent (NDA) modeling technique which is able to show behavioural intelligence. The technique used an artificial neural network to control the behaviour of a vehicle in a simple lane changing task.

Artificial neural network (ANN) models use a mathematical model for information processing with a functional architecture that resembles the neuron structure of the human brain (for an introduction to the subject see [Gurney 1997]). These models are capable of learning from training examples and demonstrating learned behaviour in unseen situations. Abdennour and Al-Ghamdi [Abdennour et al. 2006] applied ANN for the estimation of vehicle headways using data collected from different freeways in Riyadh, Saudi Arabia. Using the collected data they were able to model and train an ANN capable of estimating headways as a function of time (time series prediction) and headways as a general probability density function. Lin et al. [Lin et al. 2005] considered some sophisticated artificial neural network architectures, to model human driver behaviour in vehicle handling compared with a Driver-Vehicle-Environment (DVE) system.

The aim of the project reported in this paper was to find whether the vehicle control of the neural network based approach was an improvement over a more traditional rule-based algorithm.

2 TR Car Simulator

TRL's driving simulator uses a real Honda Civic family hatchback that has had its engine and major mechanical parts replaced by an electric motion system that drives rams attached to the axles underneath each wheel. These impart limited motion in three axes (heave, pitch, and roll) and provide the driver with an impression of the acceleration forces and vibrations that would be experienced when driving a real vehicle. All control interfaces have a realistic feel and the manual gearbox can be used in the normal manner. Surrounding the simulator vehicle are large display screens onto which are projected the images that represent the external environment to the driver. The level of environmental detail includes photo-realistic images of buildings, vehicles, signing, and markings, with terrain accurate to the camber and texture of the road surface. The driving environment is projected onto three forward screens to give the driver a 210° horizontal forward field of view whilst a rear screen provides a 60° rearward field of view, thus enabling normal use of all mirrors. Realistic engine, road, and traffic sounds complete the virtual setting. Scenario specification for the behaviour of all autonomous traffic vehicles included in simulated scenarios is determined by applying specific programming commands via SCANeR [Champion et al. 1999].

3 Development of a Neural Driver Agent

A multilayer NDA has been designed and implemented. Figure 1 shows a typical architecture of the NDA. The concept of the NDA is based on Artificial Intelligence

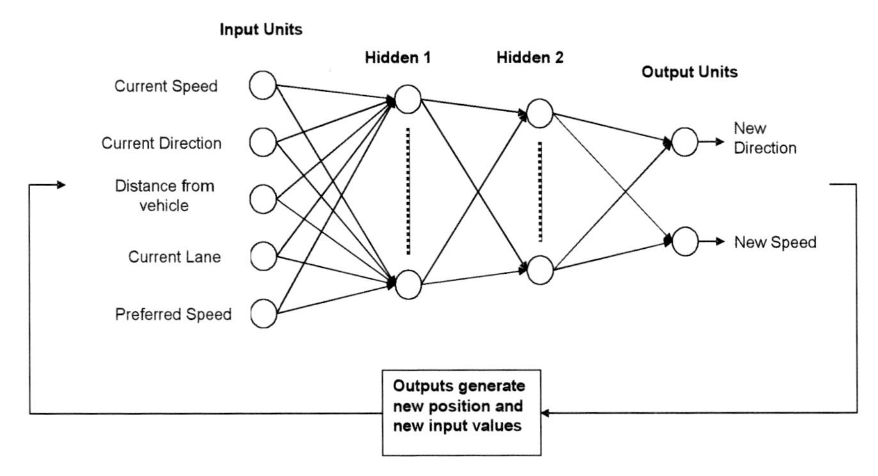

Fig. 1. Neural Driver Agent (NDA) architecture

(AI) techniques, e.g. ANN, which, at a minimum level, aims to model the system such that the system can exhibit human-like properties, for example, planning, learning, knowledge, reasoning and decision making. The NDA is based on supervised learning paradigm, with a *backpropagation* training algorithm. The inputs are propagated through the two hidden layers and output layer. The error (or mismatch) between the output and the pre-specified desired output is minimised by using a *gradient descent* rule (which essentially attempts to avoid the local minimum in training the network, by moving the weights in a direction opposite to the direction of the gradient). This allows the errors to be signalled backwards from output to input nodes until the error approaches zero. In other words, the network learns by adapting interconnecting weights.

To develop the NDA, input and output parameters to the network are defined below. Note that these parameters have been carefully selected to allow straightforward interfacing with the simulator module. A generalised expression can be derived for the multilayer NDA architecture, using gradient decent to approximate the desired output values for new direction and speed. A full derivation may be found, for example, in [Gurney 1997]. The Neural Driver Agent mathematical expressions derived were implemented in the commercial off-the-shelf NeuroSolutions software package from NeuroDimension (www.nd.com)

- Current speed, v
- Current direction, d
- Distance from vehicle, dhw
- Current Lane, l
- Preferred speed, v_p
- New direction, d_n
- New speed, v_n

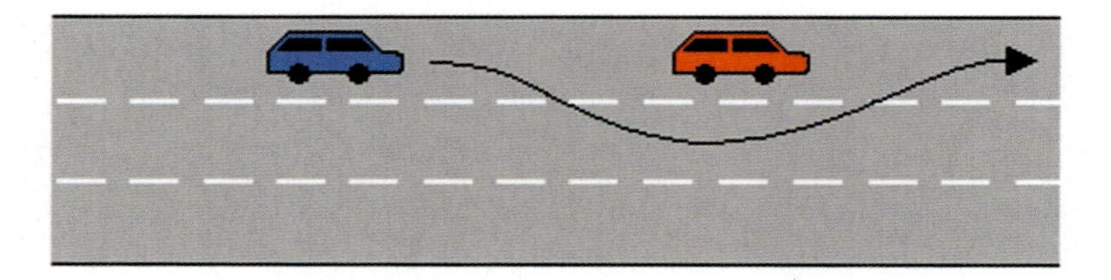

Fig. 2. Scenario set-up to generate training data for NDA

4 Results

The first part of the project was to generate the training data that would be fed to the neural network in order for the network to 'learn' how to change lanes to overtake another vehicle as illustrated in Figure 2.

Eight participants were recruited to complete a short drive on a simulated two lane motorway in the driving simulator. In this drive they were required to accelerate to a constant target speed, remaining in lane 1 of the motorway until they came across an autonomous vehicle travelling at a constant speed also in lane 1. Behaving as they would on a real UK motorway, the participant had to overtake this vehicle by moving to lane 2 and then return back to lane 1. This completed the simulator task. The data from each completed simulator run was used to train the network in how to control the driven vehicle. At each time step, the network evaluated a number of inputs, including current speed, current lane, and distance headway to the vehicle ahead, to generate outputs of desired speed and desired direction of the driven vehicle, which could be used to calculate the new position and direction of the driven vehicle. The network outputs were compared to the actual changes in speed and direction observed from the real drivers. With sufficient training the network was able to cause the driven vehicle to follow a realistic path at a suitable speed around the lead vehicle.

4.1 Comparison of Neural Drivers and Real Drivers

Figure 3 show the changes in direction produced by the Neural Driver Agent (NDA) and real drivers when performing an overtaking manoeuvre at a speed of 70 mph. The direction scale is in degrees such that 0 is straight ahead, a negative value is steering to the right and a positive value is steering to the left. The graphs show how the drivers steer to the right to move into the middle lane, then steer to the left to move back into the inside lane. The graphs show how differently real drivers perform an overtaking manoeuvre and how individual drivers also produce different behaviour at different speeds. Despite the differences in driving behaviour produced by the real drivers, the graphs show that the NDA has learnt the changes in direction required to perform an overtaking manoeuvre.

Figure 4 show the real drivers and Neural Driver Agent (NDA) accelerating to and trying to maintain a speed of 70 mph. The graph also demonstrates the differences

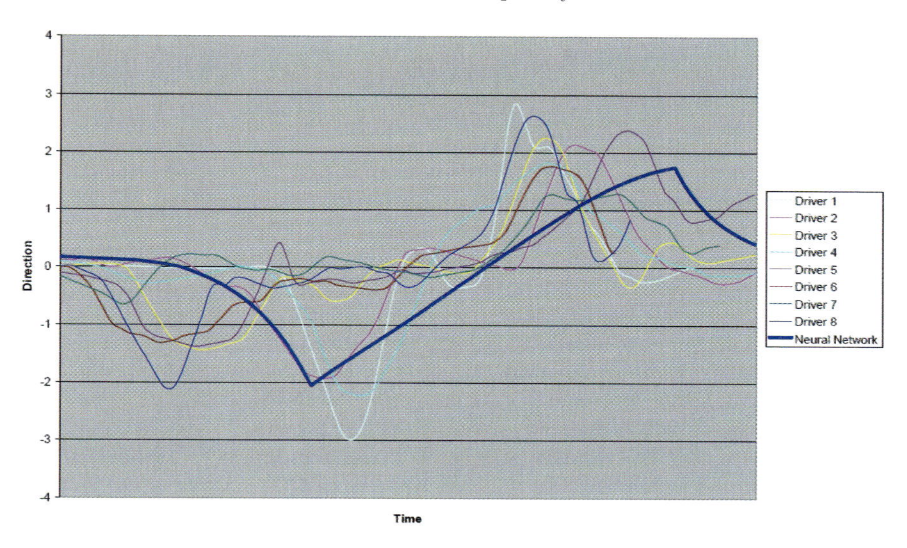

Fig. 3. Driver's change of direction when overtaking at a speed of 70 mph

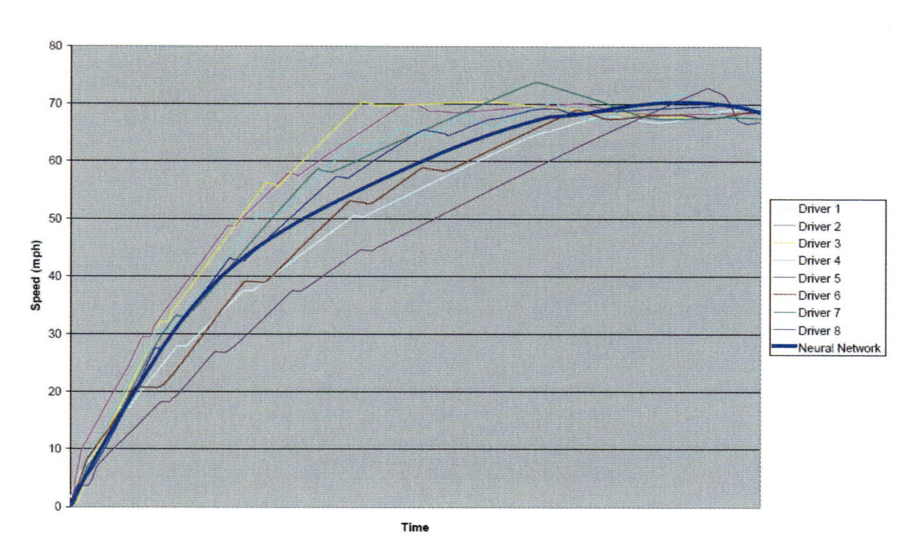

Fig. 4. Driver's speed when trying to maintain a speed of 70 mph

in the behaviours of the real drivers. When trying to achieve a speed of 70mph the NDA accelerates too much but then decelerates to maintain a speed just less than 70 mph. However, overall the NDA produces a smooth acceleration and can maintain a constant speed.

4.2 Assessing behavioural realism - results from driving simulator study

The second part of the project was to demonstrate that the neural network model was more realistic in its control of the driven vehicle than the rule-based model (from previous Loughborough University research) performing the same task. To achieve this, twelve participants were recruited to observe how each micro-simulation model controlled the driven vehicle and to rate the realism with which they thought the vehicle was being controlled. Participants each sat in the driver's seat of the simulator vehicle and were effectively 'driven' by the micro-simulation models through the simulated scenario on which the neural network model had been trained. Furthermore, participants also observed a pre-recording of how a human driver had completed the same manoeuvre. Participants were asked to rate how realistic they felt each model was on a ten-point scale from 1 to 10, where a rating of 1 indicated that they felt that the model was very unrealistic and a rating of 10 indicated that they felt that the model was very realistic. In rating the realism of the three computer models

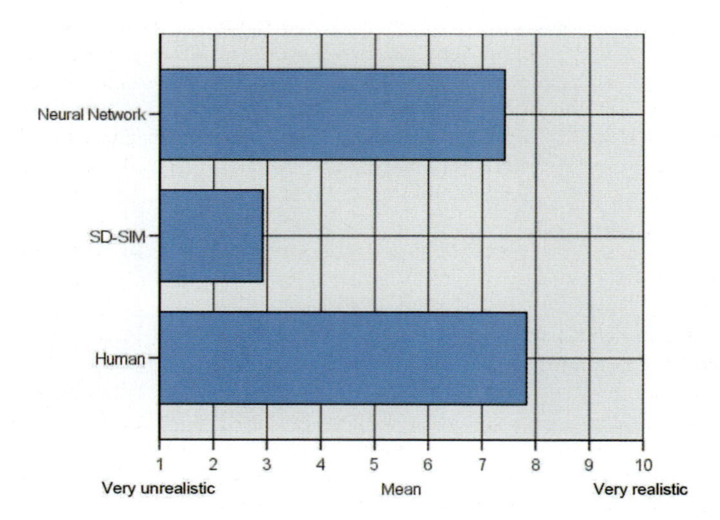

Fig. 5. Mean realism ratings of each of the models presented to participants

of driver behaviour, Figure 5 shows that on average, the human drive was the most realistic (average score of 7.83) and SD-SIM was the least realistic (average score of 2.92). It is important to note the realism of the NDA in replicating the human

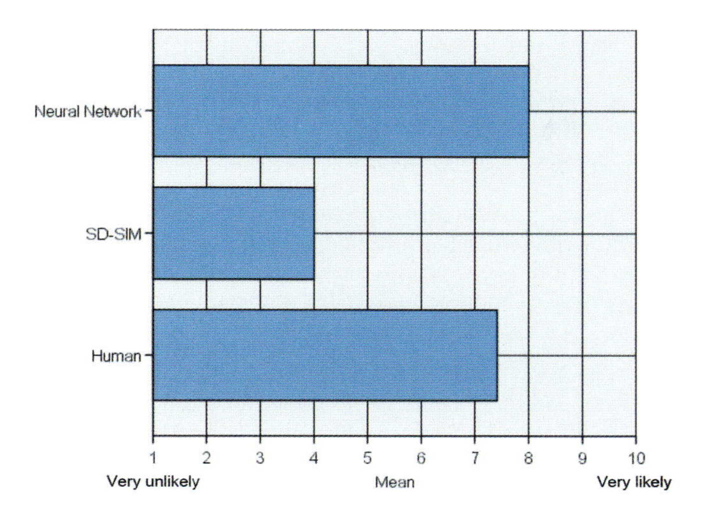

Fig. 6. Participant's mean ratings of how likely it was that each drive was actually completed by a human driver

drive (average score of 7.42). The results of the study demonstrated that participants thought that the neural network model was significantly more realistic in its control of the driven vehicle than the traditional rule-based model. Paired samples t-tests showed that the realism scores for SD-SIM differed significantly from those given for the NDA ($t(11) = 7.24; p < 0.001$) and from those given for the Human ($t(11) = 11.3; p < 0.001$), whilst the comparison of the NDA and the Human realism scores did not reach significance ($t(11) = 0.767; p = 0.46$).

To further explore the realism of the three models, participants were asked to rate how likely it was that each of the three drives presented to them was actually completed by a human driver. Again, a ten-point scale from 1 to 10 was used, where a rating of 1 indicated that they felt that it was very unlikely that the drive was completed by a human driver and a rating of 10 indicated that they felt that it was very likely that the drive was completed by a human driver. The aim was to see if participants could correctly distinguish the human drive from the SD-SIM and NDA-based drives. Figure 6 illustrates the participants' performance in classifying the models. Furthermore, participants were unable to discriminate between the human and the neural network in their control of the driven vehicle.

5 Discussion of Results and Conclusions

In general the comparison of human and neural driver agent results is good. However, there is a noticeable difference in SD-SIM and the other two models. This is contributed to by three factors: (1) fine tuning of driver characteristics in SD-SIM is currently a demanding task (2) the current vehicle model in SD-SIM lacks some of the detailed inertia and frictional effects found in the steering and suspension of real vehicles so, (3) creating a driver character in SD-SIM to match real vehicle behaviour implicitly involves some compensation for this. However, it should also be emphasised that only a small component of SD-SIM's Intelligent Virtual Driver (IVD) has been assessed. For example, the vision model which provides distance and speed estimation capabilities in lane changing behaviour was not considered in this project. Furthermore it is important to note that SD-SIM provides a framework which allows future enhancements and additions of components to improve realism. This was part of the reason for proposing the neural driver to replace some of the rule-based approaches adopted in SD-SIM. Important outcomes of the research included:

1. The study explored the use of Artificial Intelligence (AI) theories and techniques such as Artificial Neural Network (ANN) to develop a new Neural Driver Agent (NDA) model. The model was trained using captured behavioural data from participants in the simulator and demonstrated increased 'intelligence' of traffic interaction.
2. To assess behavioural intelligence and realism in driving simulation scenarios, participants rated model realism on a scale from 1-10. The results showed that human driver was the most realistic (average score of 7.83). The NDA performed well by learning and replicating human drive (average score of 7.42).
3. To further explore the realism of the models, participants were asked to rate how likely it was that each of the drives presented to them was completed by a human driver. Participants were unable to discriminate between the human and the neural network in their control of the driven vehicle.

In terms of the implications of the results in developing behavioural intelligence and realism in driving scenarios results, NDAs could potentially be developed to participate fully in driving scenarios and allow them to respond realistically in both situations to which they have been trained and novel situations. By adjusting the structure and/or connection weights of the neural network, it may also be possible to create simulated aggressive drivers, tired drivers, alcohol-impaired drivers, learner drivers, and so on. This will help to represent the range of behaviours displayed by real drivers in driving scenarios. In fact new work is underway to validate the NDA with over 50,000 individual vehicle data collected from the UK motorway.

Acknowledgment

The authors would like to thank Lena Weaver, Simulator Trial Manager for her help in conducting the simulator trials and TRL technical reviewer of the paper, Prof. Andrew Parkes for his useful comments on the paper. We are also grateful to all the participants who took part in the trials. We are particularly grateful to TRL Science and Engineering Directorate (Prof. Rod Kimber), now TRL Academy (Neil Paulley) for funding this work and to Bob Wood and John Zhao of Loughborough University for their collaboration on the project. The views expressed in this paper belong to the authors and are not necessarily those of TRL or Loughborough University.

References

[Abdennour et al. 2006] Abdennour, A.; and A.S. Al-Ghamdi (2006) *Artificial neural networks applied to the estimation of vehicle headways in freeway sections.* TEC, February 2006, 56-59.

[Allen at al. 2004] Allen, R.W.; Park, G.; Rosenthal, T.J.; and B.M. Aponso (2004) *A process for developing scenarios for driving simulations.* IMAGE 2004 Conference, Arizona, Paper N° 632.

[Allen et al. 2003] Allen, R.W.; Rosenthal, T.J.; and G. Park (2003) *Scenarios produced by procedural methods for driving research, assessment and training applications.* Driving Simulation Conference, North America (DSC-NA), Michigan, Paper N° 621.

[Champion et al. 1999] Champion, A.; Mandiau, R.; Kolski, C.; Heidet, A.; and A. Kemeny (1999) *Traffic generation with the SCANeR II simulator: towards a multi-agent architecture.* Driving Simulation Conference, DSC'99 Paris, France, 311-324.

[Dumbuya et al. 2002] Dumbuya, A.D.; Wood, R.L.; Gordon, T.J.; and P. Thomas (2002) *An agent-based traffic simulation framework to model intelligent virtual driver behaviour.* Driving Simulation Conference (DSC'02), Paris, France, September 11-13, 2002, 363-373.

[Dumbuya et al. 2003] Dumbuya, A. D.; and R.L. Wood (2003) *Visual perception modelling for intelligent virtual driver agents in synthetic driving simulation.* Journal of Experimental and Theoretical Artificial Intelligence (JETAI), 15:1, 73-102.

[Gurney 1997] Gurney, K. (1997) *An introduction to Neural Networks.* UCL Press.

[Lin et al. 2005] Lin Y.; Tang P.; Zhang, W.J.; and Q. Yu (2005) *Artificial neural network modelling of driver handling behaviour in a driver-vehicle-environment system.* International Journal of Vehicle Design, 37:1, 24-45.

[Michon 1985] Michon, J.A. (1985) *A Critical view of driver behaviour models: What do we know, what should we do?* in L. Evans and R. Schwing (eds). Human Behaviour and Traffic Safety. London: Plenum, 516-520.

[Parkes 2005] Parkes, A. M. (2005) *Improved realism and improved utility of driving simulators: are they mutually exclusive?.* HUMANIST Workshop, Conference on Application of New Technologies to Driver Training. CDV, Brno, Czech Republic, January 2005.

An Integrative Simulation Model for Project Management in Chemical Process Engineering

Bernhard Kausch, Nicole Schneider, Morten Grandt, and Christopher Schlick

Chair and Institute of Industrial Engineering and Ergonomics
RWTH Aachen University
D 52062, Aachen, Germany

Summary. The planning of development projects significantly influences the costs created by the projects as well as the success of the development projects. However much potentials are actually wasted because of the inherent complexity unmanageable for project managers. Various methods and tools, from project modeling to the workflow management system, are used to handle this complexity and to develop these potentials, but the development of software solutions alone, however, is not sufficient. Instead, an extensive instrument consisting of methods, specification techniques and software tools for the integrated transformation and simulation of a graphical process model is needed. The presented approach shows a method for the modeling and simulation of development projects in process engineering based on Petri net simulation. The simulation of an example process displays the connections between different influencing parameters such as team configuration, the availability of needed tools, the variance in processing times, and the qualification of the persons involved. Selected mathematical relationships illustrate the interaction of these influencing parameters. It could hereby be determined which parameter combination is the best to achieve setted goals like with which amount of employed staff the shortest development time can be attained. In the outlook several additional parameters are named that will be added in order to make further detailed analyses possible in a future research project.

Key words: Decision Support, Forecasting, Discrete Simulation, Simulation Interfaces, Event oriented, Work Organization, Workflow Simulation

1 Introduction

Development projects are inherently difficult to coordinate, difficult to structure and, due to the multiplicity of different participants, also hard to coordinate. The project planner has very limited control over this complexity. Currently, available tools offer few possibilities to the project manager to investigate this complexity dynamically and to accomplish "if - then" analyses. That makes the task of planning a hard to calculate risk from the point of view of the management. As a result, only 13% of work in projects in Germany is actually value-adding, resulting in a total loss of value of approximately 150 billion Euros [Gröger 2006]. Some reasons for these deficits are bad decisions in the selection of projects, yet also the insufficient defining of goals.

While these problems affect the project environment in the business there is also another area that affects the project structure. This area covers the development and continued use of findings and information in projects. This, along with the accurate implementation of employee competence and availability, must be improved through workflow planning. These problems are also well known in process engineering. The project correlations in the process development were analyzed in the Collaborative Research Center (CRC) 476. Besides these theoretical results, eight out of 12 project managers from the field of process engineering that were surveyed said that a lack of coordination and poor information flow were the main causes for sub-optimal project efficiency. Support tools were developed that improve the cooperation of the various development areas and that are meant to reduce the interface-related losses. A simulation system identifies the necessary correlations and information flows between the organizational units involved based on a semi-formal project model. The simulation clarifies the connection between the assigned resources and persons, thereby making the identification of the project duration possible through defined input of resources or vice versa. As a result, the project planner is able to analyze different workflow management structures and then plan the input of resources or the resource-relevant project structure accordingly.

2 Development Project in Process Engineering

Several example of projects were recorded that reproduce typical workflows of process engineering to take into consideration the current requirements within process engineering alongside the procedure models described in literature. The following article gives a short survey over a representative example project which was recorded in several workshops in close cooperation with experts of the chemical industry and represented an ex post view of a project finished months ago will be analized in more detail by the use of the simulation system.

For the production of a semi formal project, information that was mostly present in the mind of project leaders and coworkers in the form of experiences from past projects was first requeried, collected and documented. This seldom explicit available information goes far beyond information contained in a classical Gantt Chart representation. For the simulation, however, this source of information provides elementary data, such as alternative expiration operational sequences, qualification profiles, necessary resources availabilities as well as further conditions necessary for the execution of the activities in the form of information or predecessor activities.

The example project, the development of the synthetic material Polyamide 6 (PA6), which is usually used in the manufacturing of textiles, yet also as friction and heat resistant construction material, represents the characteristics of process engineering development processes.

Process development usually begins with literature research, which is frequently also repeated in the development project. Based on the collected information, yet also based on the experience of the developers involved, the decision for the batch or the continuous operation is made. This decision influences the additional procedure-dependent development steps. In our example, the development of the PA6 process was performed in cooperation with chemical engineering companies; the developments of the reaction, separation and extrusion follow. These developments result individually, yet depend heavily on each other, founding the basis of the complexity within the development projects of chemical engineering. The development of the facility area necessary for the various steps is based on the representation of the mathematical, chemical and physical correlations. Consequently, the main task is the creation and analysis as well as the improvement of these models. To conclude, the final decision regarding the plant concept is made based on the simulation results of previous work steps.

2.1 Definition of the Simulation Approach

A workflow simulation model of development projects in the chemical industry was developed at our institute in recent years. One way to differentiate simulation models is by the level of detail found in human modeling. VDI-Guideline 3633 distinguishes between person-integrated models (person as reactive action model) and person-oriented models (display of various additional traits possessed by person) [VDI 2001, Zülch et al. 2004]. Furthermore, simulation models of product development processes can, similar to VDI-Guideline 3633, be differentiated by two forms of model logic:

1. In actor-oriented simulation models, system dynamics are produced by actors (modeled persons or organizational units) based on specific tasks [Steidel 1994, Christiansen 1993, Cohen 1992, Jin et al. 1996, Levitt et al. 1999, Licht et al. 2004].
2. In process-oriented simulation models, system dynamics are produced by activities through the usage of resources (persons, tools) [Browning et al. 2000, Cho et al. 2001]. According to this terminology, the workflow simulation model in process engineering that will be presented here can be characterized as a person-oriented and process-oriented approach.

3 Existing Approaches

In the field of process and product development processes only a few adequate simulation techniques are well established. The so-called Virtual Design Team (VDT) is an actor-oriented model especially designed for the simulation of product development projects which was created by Levitt's research group at Stanford University. Early versions of the VDT were already able to model actors and tasks, as well as the information flow between these two [Christiansen 1993, Cohen 1992]. Subsequent versions then also took into consideration the different goals of actors, the construction of exceptions, and in addition, exception handling [Jin et al. 1996, Levitt et al. 1994, Levitt et al. 1999]. A process engineering context is not considered in this model, and participative creation of the simulation model or optimization of workflow management will not be supported through the methodology.

Independent of Levitt's group, Steidel managed to develop a further actor-oriented simulation model for product development processes [Steidel 1994]. This model also ignores particularities of process engineering. Likewise, participative creation of the simulation model or optimization of workflow management will also not be supported through the methodology.

Raupach formulated a process-oriented approach for the simulation of product development processes so that consistency can be observed in various construction solutions. The product structure is accounted for in great detail through this approach [Raupach 1999]. This fact makes it hard to apply in contents with inherent variability, e.g., the process engineering context, participative process creation, and optimization of workflow management. These points, as well as interdependencies between project success criteria and factors influenceable by technical planning, will not be examined in this approach.

Aside from the complexity of development projects, the meeting of decisions at times of uncertainty is an important aspect of simulation. Krause uses colored Petri nets in combination with stochastic procedures in order to sufficiently depict these decisions during simulation [Krause et al. 2004]. The planner first roughly models the activities of the development process; these activities are then further specified during the simulation run through the accessing of a library. The dynamic calculation of the model structure at cycle time adequately depicts the uncertainty-afflicted cycle of planning processes. However, participative modeling and optimization of the processes according to defined restrictions and target criteria is missing.

Eppinger's research group at the Massachusetts Institute of Technology developed numerous process-oriented simulation models [Browning et al. 2000, Cho et al. 2001, Cho et al. 2005].

Browning's simulation model assumes that an unlimited supply of resources (in this case, employees) exists, meaning the simulation results of this model are limited in their representation of reality. Cho's simulation model does take note of the limitation of resources available in a product development project, yet a corresponding processing of multiple activities is also not possible in this case. An organizational connection to process engineering is non-existent, and participative process creation or an optimization of workflow management is not intended. Interdependencies between project success criteria and factors influenceable by technical planning will hardly be considered.

A process-oriented model for the simulation of a factory-planning project was developed by a research group headed by Tommelein at the University of California at Berkeley [Gil et al. 2001]. This model observes the effects of altered requirements on the planning process and the project length of construction projects. Particularly, examination of so-called postponement strategies occurs, in which the start of a succeeding operation is purposely delayed in order to increase the quality of the work results of the preceding operation. Similarly, the simulation model assumes an unlimited supply of resources. However, in a process engineering context, participative process creation or an optimization of workflow management is not dealt with. Interdependencies between the technical planning of influenceable factors and project success criteria are not sufficiently taken into consideration in this model.

According to our requirements, the person-centered simulation model of Licht [Licht et al. 2004] offers a more suitable approach to analyzing development processes of products and processes. The model includes many different process specific aspects of the process, such as type and complexity of products, characteristics of the employees, tools, organizational structure, etc. Due to the person-oriented approach, the model also serves as a realistic method for employee management by providing employees' behavior. The negative consequence, however, is that the model is very complex and therefore difficult to apply.

4 Integrative Simulation Model

The simulation system presented here offers a suitable technique for project planners in order to compare several alternative ways of project organization at an early stage, with respect to the number of persons, tools, time and other resources involved.

The process of analyzing different organizational project structures including different configuration of resources starts with a semi-formal modeling of the project structure including among other things the fuzzy coherences between the actors performing the activities, between the activities and their duration. Especially developed for the modeling of weakly structured and highly communicative and coordinative development processes, the modeling language C3 consists of 14 base elements. Figure 1 shows the besides the symbol of the starting and the end node the nine most important elements in a short section of a chemical engineering model.

The logic or temporal coherences between different activities is described by the control flow. The flow of information elements describes the emitter, the receiver and the type of information. The special characteristic of the synchronous communication is that all involved actors and resources must be available at the same time. If not, the synchronous communication cannot be executed. This possibly affects many other activities and other parallel control flows by resources or actors tied up for the communication for a certain time. The splitting note finally splits information. Therefore, a parallel execution of activities is possible. The synchronization node merges parallel activities.

With its close connection to the easy to understand and semi-formal modeling language C3 [Killich et al. 1999], designed at our institute and described in more detail in Kausch [Kausch et al. 2007], the simulation model is concise and easy for the user to apply. The goal of this simulation model is to combine the advantages of C3 [Eggersmann 2004, Schneider et al 2006] and the advantages of the simulation, that is to say, the possibility of planning, analyzing and rearranging the development process based on mathematical constraints. In addition, the model used offers the chance to optimize the development process with respect to the development

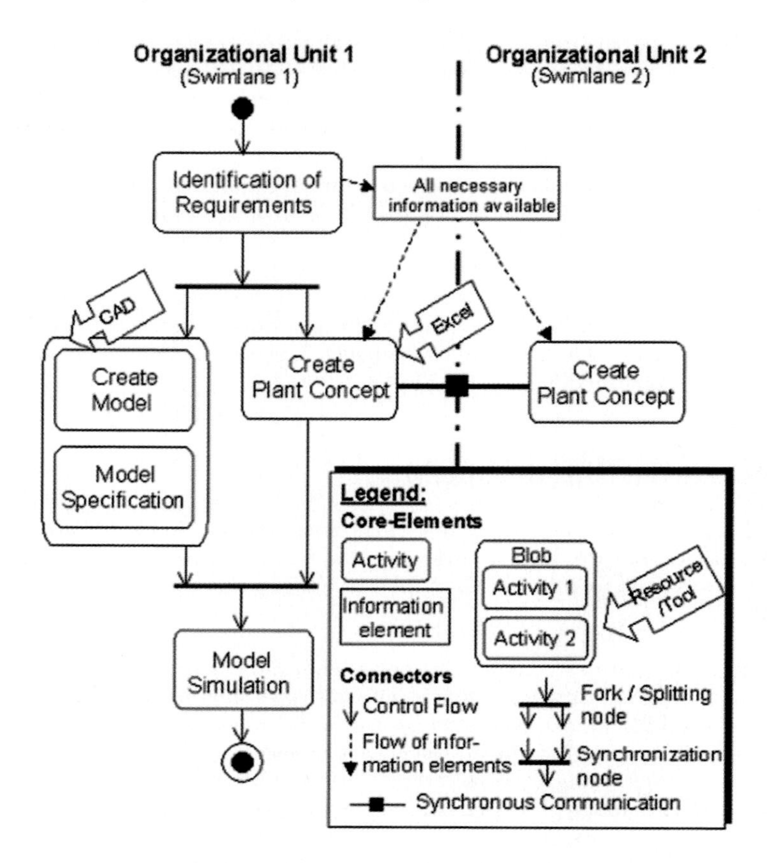

Fig. 1. Core elements of C3 modeling language

duration as well as in consideration of resources and the development costs. The entire simulation model is based on the following five partial models: 1.) the task network, 2.) the task, 3.) the employee, 4.) the work tool, and 5.) the information, which will be examined in greater detail in the following.

4.1 Task Network Model

The development of a new or modified chemical process usually takes place in team spanning development projects. It is in these projects that the complexity concerning the organizational structure as well as the workflows should be reduced. The model concept of the task network describes the workflow management of the development project. In addition, the individual phases of the development process will be divided into work tasks through the use of a workflow plan (a so-called task network). Predecessor-successor-relationships, i.e., the logical order of execution - for example, due to causal relationships between individual underlying activities - of the tasks will be laid down in the task network. It is hereby determined that the literature research precedes the additional analysis. The workflow plan is primarily participatively recorded and displayed through the C3 modeling language. The work tasks of the task network are assigned to organizational units for execution. Apart from the chronological sequence of tasks, the assignment of work equipment for the associated tasks is also displayed in the task network. An overview of the PA6 development process, described earlier, with 79 activities is schematically displayed in Figure 2.

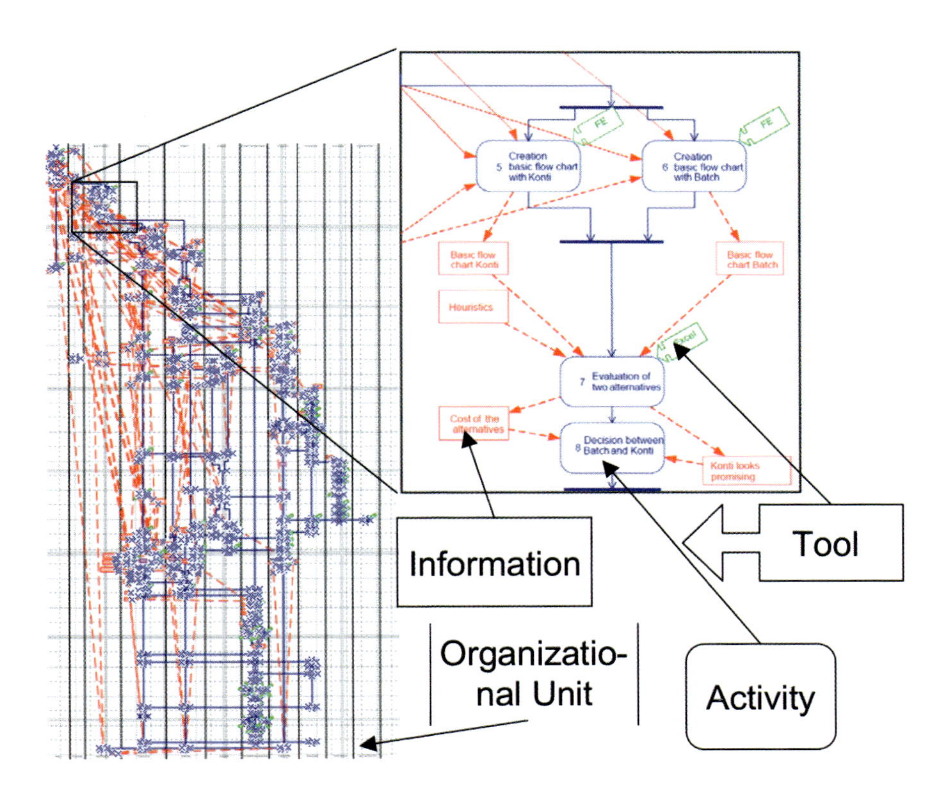

Fig. 2. Schematic illustration of the PA6 development Process including a detailed view of some basic C3 elements

Additionally, in the upper right excerpt, a detailed view into the process is given, where the main elements of C3 are marked and briefly explained in context. A software environment, especially designed in the research project to support recording and visualization of work processes with the C3 method, supports the recording process as well as the visualization and software based transformation of the working process structure. Schneider and Gerhards [Schneider et al. 2003] describe this software environment for work process modeling in more detail.

4.2 Task Model

The task network consists of the tasks in the development process that need to be worked on. The processing of each individual task is described in detail in this model concept. Within the tasks there is information about the subject matter needing to be processed, a necessary work tool, a profile of possible persons to do this processing, input and output information of the task as well as the expected duration needed to process the task. For the processing of a task, a qualified person and, if necessary, adequate tools are selected to achieve the goal of only implementing the most qualified employee actually available for the handling of the task. Each person is then also assigned a value that reflects the quality of the person, dependent on the task at hand and the required tools to complete the task. This value is calculated from the weighted sum of the person's assigned characteristics (see Model Concept of the Employee).

The weighting and the different attributes are not constant and can be varied depending on the area of application. The most highly qualified person will then take on the task, though it may occasionally be the case that the basic skills needed for a certain task are not possessed by anyone. In such an event, the task cannot be completed until someone suitable for the task becomes available. Only once the adequate labor and essential work tools are available can the task be carried out according to its duration, which depends on the underlying distribution function and the person employed for the task.

4.3 Employee Model

According to the person-oriented basic approach of simulation, the definition of the characteristics of employees (participants) in this model concept is of particular importance. At the same time, an attempt is made to model the person as realistically as possible. This entails displaying employees' characteristics and abilities that have an influence on the allocation of persons to the various tasks as well as the task processing time and work quality of the different development process tasks. The described attributes of an employee are summarized in the following:

- Productivity of an Employee
  Each person is assigned a numerical value that describes the individual productivity, i.e., output. This value improves the quality of the employee in the selection of the most qualified employee for a task, and also has an influence on the processing time of a task.
- Qualification in terms of a particular area of work
  The tasks of the development process are arranged into swim lanes in accordance with C3 modeling. These swim lanes describe the areas of work, for example, such as in the PA6 Processes case study in which the work areas of Simulation or Separation were described. The persons possess abilities and skills that qualify them for the processing of tasks in certain areas of work, yet then also make them unsuitable for others.
- Ability to deal with particular work tool
  Several tasks require a work tool such as a software tool or a machine for their processing. The persons possess abilities and qualifications that describe how

well they can handle certain work tools. This means a person must not only bear the appropriate qualifications to complete the task, but they must also have the ability to carry out the task through use of the necessary work tools.

- Learning aptitude
  An employee begins a career with certain basic qualifications, i.e., abilities that were acquired during schooling, or inherent characteristics. During the course of a career, however, a person's abilities can change. Due to routine tasks and new methods and expertise, certain qualifications can actually be improved. Alternatively, abilities not put to use over a greater period of time can also be weakened. This capacity to learn and unlearn is shown in a simulation model through a learning curve that is attributed to each person. A more detailed description of the learning curve will be presented later on.

Personal qualifications and abilities are taken into account in the model concept in terms of recognizing that each person is able to act out a variety of activities. This portfolio of possible activities can be directed at specific job descriptions that are representative of the different organizational units and work means related to the process.

4.4 Work Tool Model

The influence of work tools on the completion of tasks by an employee is held in the partial model of work tools. The allocation of work tools to tasks results through the work organization of the development project. Simultaneously, the information of which work tools can be used for which task is already retained in the model of the task network. Due to their scarcity, work tools must be reserved prior to their use. Also, a tool can be used by only one employee at a time, though more than one tool can be used for a specific task. The amount of possible work tools cannot be exhaustively declared since the amount of possible tasks in need of completion, detached from individual case examples, cannot be fully indicated. Thus, similar to the task network and the work organization in relation to the development project that is to be simulated, the list of work tools must be created and must be specific. The level of detail is also to be specified individually for each case. This means that it may be enough in a project to simply differentiate between work tools for the creation of technical drawings between drawing board and CAD; in other projects, due to the use of varying computing systems and thereby related file formats, there must be distinction between different computing systems.

4.5 Information Model

Information should be viewed in the same light as work tools. Information is already assigned to tasks in the task network and has an influence on the duration of the development project. Information can be grouped into input and output information. Input information describes files or documents that are necessary for the processing of a task. The processing of a task cannot start without this information.

For example, for task seven of the case example (see Figure 1), evaluation of two alternatives for the creation of a basic flow chart with Batch or Konti requires information about various heuristics as well as output from the basic flow charts of

Batch and Konti. These can either be produced in the form of output information through a different task, such as the Batch or Konti information which is linked to the previous tasks, or be made available outside of the analyzed workflow as in the case of the heuristics. Through the processing of a task, output information is treated as its result. The results of a task, which may eventually be needed for the processing of later tasks, are described and are made available as input information.

5 Implementation of the Simulation Model

To show the implementation of the simulation model, the Polyamide 6 process (Eggersmann 2004) was used as an example case of the CRC. The underlying process here, consisting of 79 activities executed by the coordination between eight organizational units (separated by swim lanes in the C3 model), describes the different phases of new development for the manufacturing of PA6.

To maintain the distinctiveness of the C3 language the simulation model was implemented using a person-oriented and process-oriented approach. Also, to formally describe the simulation model, the notation of Timed Stochastic Colored Petri nets was taken up. The development project was mapped into a directed graph consisting of places, transitions, arcs, and markings. A great advantage of this simulation notation is that a stepwise simulation can easily identify weak points. In this case, Petri net tokens as representatives for active elements indicate the status of work progress as a result of possible weak points.

The simulation model was implemented using the Petri net simulator Renew [Kummer et al. 2004]. Renew is a Java-based high-level Petri net simulator developed at the Department of Informatics at the University of Hamburg. The simulation tool provides a flexible modeling approach based on reference nets as well as a user-friendly design by the use of a graphical presentation. Renew is a computer tool that supports the development and execution of object-oriented Petri nets, which include net instances, synchronous channels, and seamless Java integration for easy modeling.

The entire Petri net model according to the description of the Polyamide process is composed of different sub-networks that correspond to partial models that, for instance, represent the universal model. The implementation of this partial model in the form of sub-networks will be examined more closely in the following.

5.1 Task Network

The Task Network describes the workflow management of the development project. The predecessor-successor-relationships between individual tasks are defined in the corresponding Petri net. Certain tasks are released for further processing through appropriate transitions in this network when all necessary predecessor tasks have been completed and the adequate persons as well as resources (work tools, input information) for the processing are available. A section of the task network of the PA6 Process is displayed in Figure 2. Based on the process-oriented approach, the task network builds the link between the partial models. Here are the rough correlations, such as how the development project uses workflow management and the necessary resources for the processing of individual tasks, whereby the exact processing of tasks are represented in the network of the task.

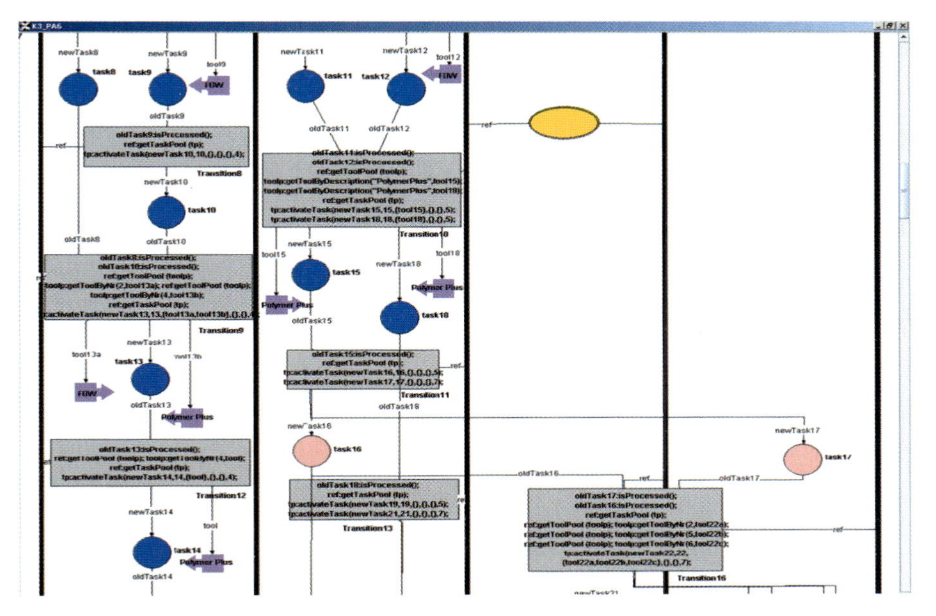

Fig. 3. Screenshot of a section of the PA6 process

5.2 Task

The net for the representation of the processing of a task builds the link between the partial model of the work tool and the employee. Here, the person who will process the task is chosen and the necessary resources are reserved. In doing so, the basic conditions are directed at the person who is qualified for the processing of the task. These requirements are implemented in the respective task and organized according to the area of application, with the most qualified person executing the task. The qualification level (Q_L) is calculated as follows:

$$Q_L = \alpha P + \beta Q_w + \gamma Q_t \tag{1}$$

The weights α, β and γ determine how strong the influence of an attribute is on the quality level of a person. These three multipliers together describe 100% of the influence. According to the model concept of the person, the attributes productivity P, qualification based on the field of work Qw and the ability and qualification to handle a work tool Qt are viewed as influencing variables. Moreover, the duration of a task is determined and thus processed in the network of the task. Effort and duration for the processing of a task depend on the estimated average processing duration as well as the qualification and proficiency level of the specific employee. The choice of work tools used along with the procurement of additional information can also have an effect on the duration and processing of a task. In order to realistically depict the processing time of a task, which can only be approximated, the aid of a probability distribution is employed. A normal distribution with relative variance between 10% and 30% of the mean was established for the first test runs of the simulation model. The administration of the tasks of the workflow is implemented in the Task Pool. The Task Pool is a help network that, in combination with the Task Net, displays a task on the model concept. The various tasks are initialized and managed in the Task Pool.

5.3 Person

The employees involved in the project, inclusive of their characteristics and capabilities, are implemented in the Person Net. The management of employees is organized in an auxiliary net, the so-called Person Pool. Here, the current number of available persons as well as their current status - "currently in processing" or "free for the next available task" - is deposited. Before a task can be processed, however, a search occurs in the net for the fitting employee for the processing of the task. (see the Net of a Task).

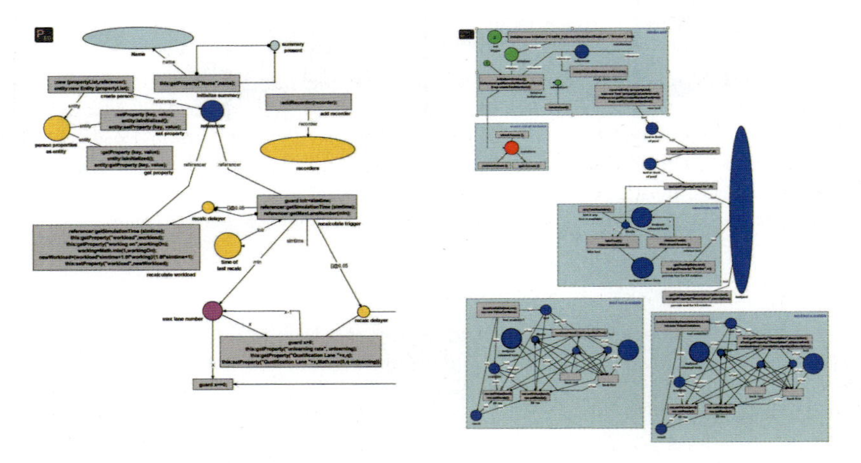

Fig. 4. Screenshot of the Renew Person Net and the Person Pool

In Figure 4 a section of a screenshot of the Person Net as well as the Person Pool is mapped. Task-specific abilities of a person are improved, thereby increasing the attributes of that person when a task is processed. This learning ability of the employee is implemented through a learning curve as follows in Figure 5.

In the implementation of the learning curve it was possible to deal with the learning ability of every single employee. Therefore employee-specific factors (a, b, c) were utilized. They describe the individual learning aptitude like the receptivity, the

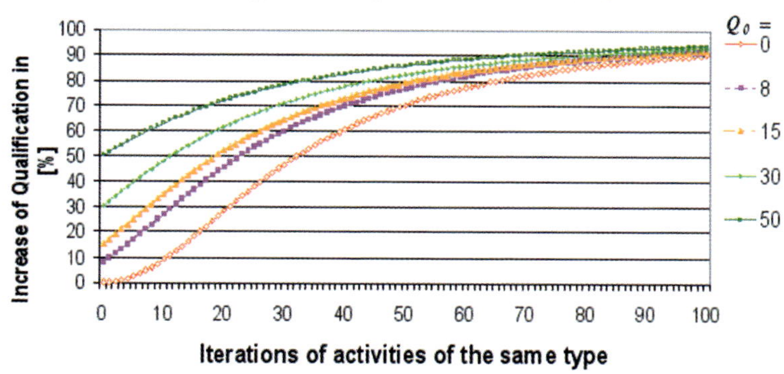

Fig. 5. Coherence between individual qualification and number of task executions

memory and the neglect of different persons. The basis-qualification at the beginning of the project is Q_0.

5.4 The Tool Net

The work tools available for the work process are administered in the Tool Net and the Tool Pool. In the model presented here, a name and a distinct identifier are sufficient as a characteristic of a work tool. Modeled on the Person Pool, the Tool Pool implements the maintenance of work tools, i.e., the current number as well as status of available work tools is accounted for. It may sometimes be the case that another person is already using a specific tool, leading to waiting times.

5.5 Implementation of Additional Functions

A universal model composed of further help networks exists in addition to the networks that describe partial models. In this universal model functions, such as the initialization of the model or the output of simulation model results, are implemented. These act as links between the various nets. The input data of the simulation model (the description of the tasks in the development project, including their demands of the employee as well as their necessary resources, the amount and attributes of the employees involved in the project, as well as the work tools available for the project) is organized in tables and can be viewed with the help of the initialization network. Additional functions, for example, the calculation of the normal distribution of the processing time or the printout of simulation results, are implemented in independent Java classes whose functions are invoked and performed in corresponding parts of the network, more precisely, in the transitions.

6 Results and Discussion

Before the models can be used to identify causes and effects [Daalen et al. 1999] the models must be checked to see if they are valid representations of the systems to be studied. VDI 3363 [VDI 2001] additionally suggests the comparison of real data to simulation results. The C3 modeling method mentioned above has been used for the assessment and modeling of different development processes in chemical engineering, performed in cooperation with experts and researchers from this field. Concerning the structural validation of the simulation model, the coordination of the numerous individual parameters among each other should be seen as particularly critical. These parameters produce extremely complex system dynamics through which the investigation and evaluation of the models is in turn made more difficult.

In the first test runs of simulation the number of persons was varied. Following the valid results of this pre-test the number of tools was also varied. The influence of these factors on the simulation time was then examined in order to judge the validity of the simulation model. To do so, the expected durations of the individual tasks were acquired in multiple expert workshops. As described in the following, these initial test runs showed satisfactory behavior.

6.1 Examination of Dependence between Number of Employees and Total Project Duration

The relationship between the total duration of the development project and the number of organizational units working on them - in the present case identical to the persons working on the task - was analyzed in the first simulation runs. Also, it was assumed in the form of the simplest case, that only one person processed a task. For this comparison the amount of persons was varied between one and 11. The variance of the expected duration was still regarded as an independent variable and then changed in three steps, between 10%, 20% and 30% of the mean, so that ten (n=10) runs are simulated for each of the possible 33 (b=33 out of: 11 differing amounts of people x 3 differing variances) combinations of variables. The corresponding hypothesis states that the duration decreases with each additional employee. Experts forecast that the influence of the number of employees will far exceed the boundary-defined duration variable. Therefore, it was the total duration, forecasted through the simulation model, which was to be analyzed.

First results (see Figure 6) were first examined on the basis of significant differences in duration. Through a one-way analysis of variance (ANOVA, $\alpha = 0.05$) it is shown that there is no significant difference within the groups that have the same number of employees. This confirms the hypothesis that with any of the possible deviations from the expected value of duration time (between 10% and 30%) that are regarded in the simulation, no significant change in total duration time takes place. This is supported independently of the predicted duration and describes a balancing effect on the variance of a large number of activities (a=79). Experience shows, however, that projects usually do encounter delays, which is why the variance in the more detailed simulation must be extended by a right-skewed b-distribution.

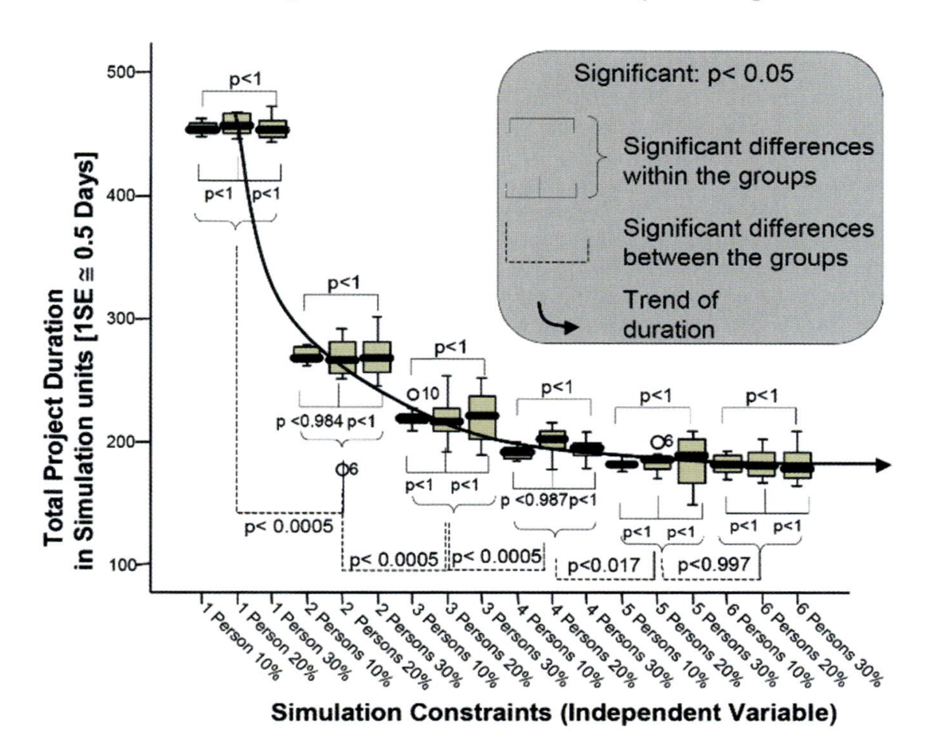

Fig. 6. Overview of the dependencies between the total project duration and the number of employees and the variance of the expected value of activity duration time

The simulation further shows that the duration can be reduced by approximately 60% through employment of more than five persons. After employment of more than six persons though, no significant reduction in duration can be measured. This is due to the project structure's task network in which no more than five tasks can be carried out at the same time, thereby also not being able to be processed by more employees. Task sharing within a task is promoted through this, and the resulting implications were examined in further studies. Subsequent reasons for an unwanted short duration through a high number of employed organizational units lie in synchronous communications. These occur in specific intervals, lying between the tasks, and thereby occupy the required persons of the participating organizational units. In doing so, the employees are picked from the task network and "scheduled" for the discussion through the simulation. These employees can process no other tasks during this time. These communication relationships are a particular feature of development processes so that the high significance assigned to them through the simulation corresponds to actual conditions. After this short pre-test further siumulation runs were conducted to allow a more detailed analyzes.

6.2 Parameterization

To determine the best constellation for the realization of the project, the numbers of persons and tools were systematically varied. At least nine different tools and two different actors are necessary to conduct the project. This "basis" constellation is then extended, first by additional highly qualified actors able to conduct each activity, and second, by a multiplication of tools. The number of working persons (N_{WP}) and the total number of tools (TN_{OT}) were systematically multiplied so the different constellations shown in Table 1 were accomplished with partly different number of iterations mentioned in the cells with n_{121}, n_{221}, etc.

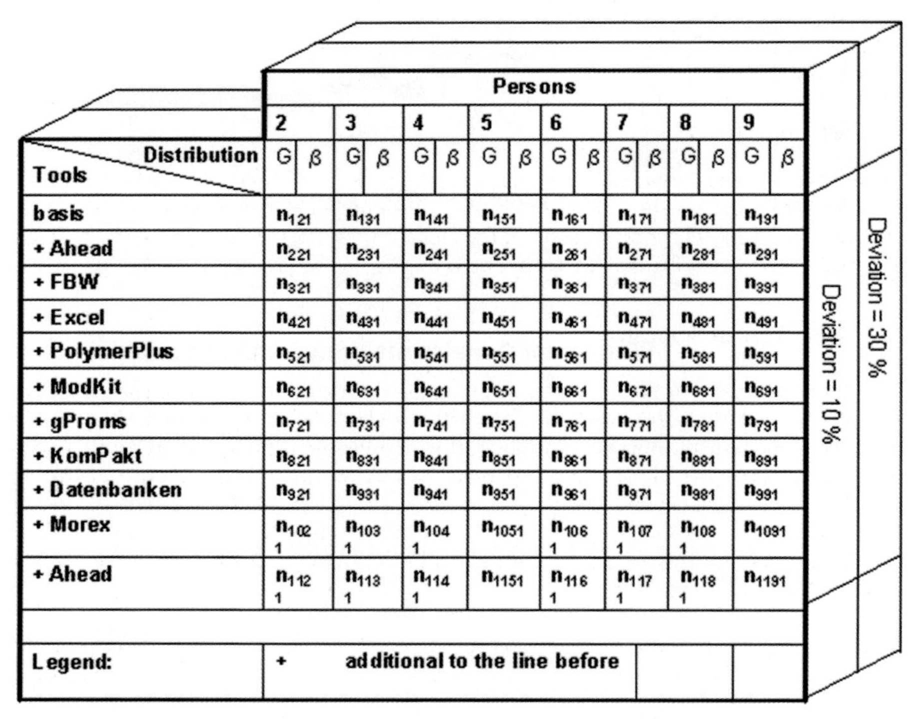

	Persons							
	2	3	4	5	6	7	8	9
Tools / Distribution	G ß	G ß	G ß	G ß	G ß	G ß	G ß	G ß
basis	n_{121}	n_{131}	n_{141}	n_{151}	n_{161}	n_{171}	n_{181}	n_{191}
+ Ahead	n_{221}	n_{231}	n_{241}	n_{251}	n_{261}	n_{271}	n_{281}	n_{291}
+ FBW	n_{321}	n_{331}	n_{341}	n_{351}	n_{361}	n_{371}	n_{381}	n_{391}
+ Excel	n_{421}	n_{431}	n_{441}	n_{451}	n_{461}	n_{471}	n_{481}	n_{491}
+ PolymerPlus	n_{521}	n_{531}	n_{541}	n_{551}	n_{561}	n_{571}	n_{581}	n_{591}
+ ModKit	n_{621}	n_{631}	n_{641}	n_{651}	n_{661}	n_{671}	n_{681}	n_{691}
+ gProms	n_{721}	n_{731}	n_{741}	n_{751}	n_{761}	n_{771}	n_{781}	n_{791}
+ KomPakt	n_{821}	n_{831}	n_{841}	n_{851}	n_{861}	n_{871}	n_{881}	n_{891}
+ Datenbanken	n_{921}	n_{931}	n_{941}	n_{951}	n_{961}	n_{971}	n_{981}	n_{991}
+ Morex	n_{1021}	n_{1031}	n_{1041}	n_{1051}	n_{1061}	n_{1071}	n_{1081}	n_{1091}
+ Ahead	n_{1121}	n_{1131}	n_{1141}	n_{1151}	n_{1161}	n_{1171}	n_{1181}	n_{1191}
Legend:	+ additional to the line before							

Table 1. Four dimensions of independent parameters

Additionally, the statistical dispersion (variance of task duration: V_{TD}) was varied between 10% and 30% with the Gaussian normal distribution (G) and with the right skewed beta (b) distribution ($\alpha = 6, \beta = 3$) (variation of the distribution: V_{DT}). As a consequence of this distribution, combined with the abstract elements in the process flow, several thousand possible sequences arise. In total, 5222 runs with 155 selected different constellations of independent variables were performed.

6.3 Dependent Variables and Hypotheses

The most important dependent variables are total time of project duration (TT_{PD}) and total time of work (TT_W), also seen as total effort needed to execute project. TT_{PD} describes the duration of the project from the beginning of the first task to the end of the last task. The dependencies are the level of parallelization (LP) and the grade of integration of the persons available (G_{IP0}). The average of the personnel workload is the sum of G_{IP} divided by the N_{WP} and considered as the average integration of working persons AG_IP. The following null hypotheses were generated as key hypotheses for further analyses:

H01: The total time of project duration (TT_{PD}) is independent of the number of working persons (N_{WP}).
H02: The total time of project duration (TT_{PD}) is independent of the total number of tools (TN_{OT}).
H03: The total time of project duration (TT_{PD}) is independent of the variance of task duration (V_{TD}).

By analyzing these coherences, prognoses can be made about the cost of the developing process, the realistic position of milestones and the consideration of risk management, the workload of actors and tools, the time of the assignment of actors and the application of tools and other resources.

6.4 Simulation Results

The first overview can be gained by graphically analyzing the effect of the variation of the most cost-intensive variable N_{WP}. On a 5% level of significance ($\alpha = .05$), using the confidence intervals of 95%, highly significant (**, $p \leq .0001$) differences between the groups of 2, 3, 4 and 6 persons can be discovered (see Figure 7). Additional the percentaul reduction of TT_{PD} is given in the sense of sensitivity analysis. An increase above at least 6 persons does not significantly affect the TT_{PD}, though this is the one-dimensional view to the simulation result. Additionally, the other independent variables must be taken into account.

A more detailed high dimensional five-way analysis of variance (ANOVA) permits the analysis of the main and the side effects. As already seen, the most important variables are N_{WP} and TN_{OT}. As an example, Table 2 shows the results of the ANOVA for TT_{PD}:

It can be summarized that H_{01}, H_{02} and H_{03} must be revoked. However, the ANOVA cannot explain which factor affects the dependent variable and to what extent a (estimated) measure of effect strength $\hat{\omega}^2$ expresses the portion of total variance explained by a single (statistically significant) effect. It can be calculated as follows [Hays 1973, Eimer 1978, Bortz 1999]:

$$\hat{\omega}^2_{Effect} = \frac{QS_{Effect} - df_{Effect} \cdot MQ_{Error}}{QS_{Total} + MQ_{Error}} \tag{2}$$

$$MQ_{Error} = \frac{QS_{Error}}{df_{Error}} \tag{3}$$

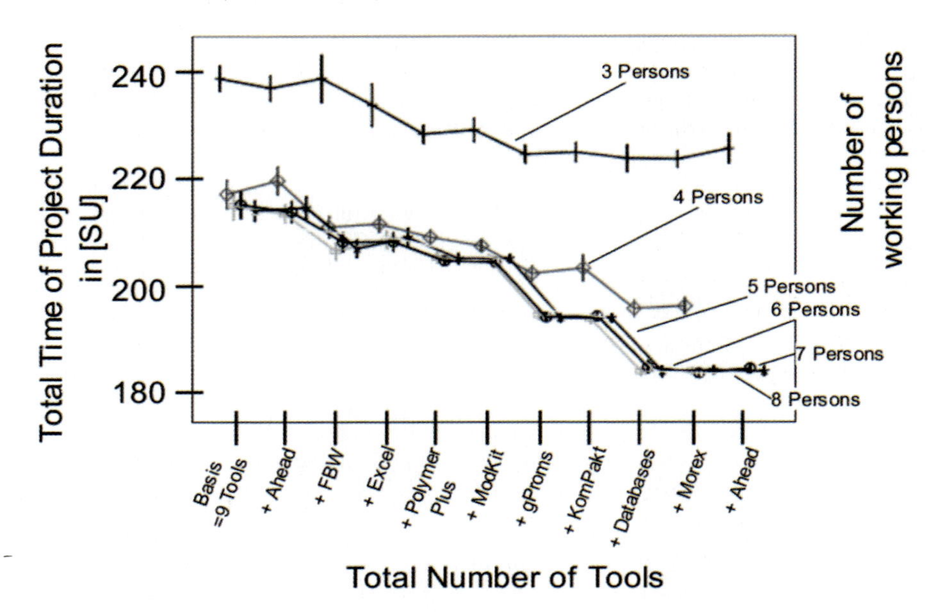

Fig. 7. Error bars (95% CI) showing the interdependencies between N_{WP}, TN_{OT} and TT_{PD}

Effects	df	F	p
N_{WP}	7	3054.85	0.000
TN_{OT}	19	336.78	0.000
V_{TD}	1	89.98	0.000
V_{DT}	1	640.81	0.000
$N_{WP} \times TN_{OT}$	90	8.31	0.000
$N_{WP} \times V_{TD}$	7	5.83	0.000
$TN_{OT} \times V_{DT}$	2	42.23	0.000
$N_{WP} \times V_{TD}$	7	3.73	0.000
$N_{WP} \times TN_{OT} \times V_{TD}$	5	4.56	0.000

Table 2. 5-way ANOVA of TT_{PD}

The main effect strengths occur as follows: 69% of the variance of TT_{PD} can be explained by variation of N_{WP}, 15% by variation of TN_{OT} and between 1.5% to 2.0% with the distribution, the variance V_{TD} and the effect of interaction between $N_{WP} \times TN_{OT}$. The LP is explained by 57% with the N_{WP}, 23% with TN_{OT} and with 3.2% with the effect of interaction between $N_{WP} \times TN_{OT}$. AG_{IP} is highly affected by N_{WP}: 99% of the effect detected on this variable can be explained by the variation of N_{WP}. In conclusion, the most important variables are N_{WP} and TN_{OT}.

The analysis of sensitivity is more detailed in terms of optimizing the project progression, taking into consideration necessary personnel and financial effort. As shown in Fig.5, 6 is the optimal number of persons involved in the project. With increasing personnel assignment between 2 and 6 persons the total time of project duration could be decreased by more than 36% or 89 [SU]. With 1[SU] 0,5d the total project duration could be reduced more than 2 months.

6.5 Support in Practice

The interpretation of the simulation results requires a high level of expertise in simulation and data analysis. To provide adequate support for experienced project planners, the simulation results can be transformed into a graphic visualization similar to common project plans, showing the sequence and coherences of activities.

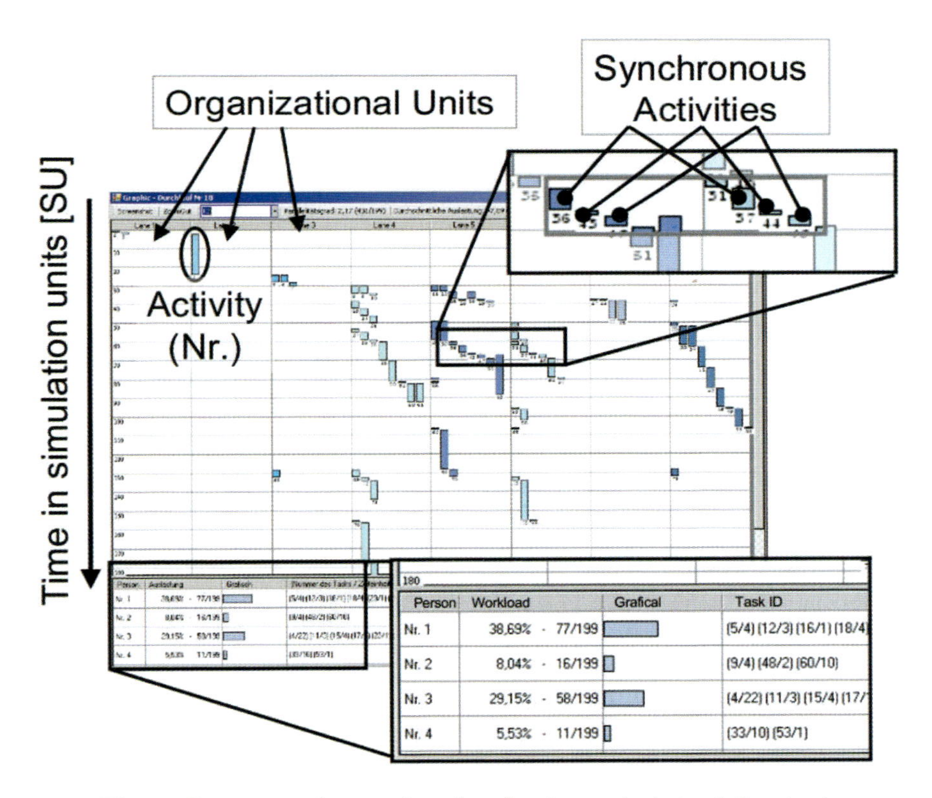

Fig. 8. Prototype of a user interface for the project simulation tool

Figure 8 shows the prototype of a user interface where key elements can be detected and the project plan can be developed from. Colored bars and the activity ID numbers indicate the point in time where the activity starts and ends. Additionally, the organizational unit responsible for the execution of the activities, special coherences such as synchronous communication between certain activities, and the actor who executes the activity can be analyzed. Finally, different information about the absolute and relative workload of each actor as well as the variables mentioned before (TT_{PD}, L_P, G_{IP}) is displayed. The possibility of gaining a detailed overview of estimated project progression depending on team constellation, appropriateness of the actors, resources and tools, and - with regard to the risk management - the variation of the duration of tasks and activities, leads to better overall support of project planners. The recent enhancement of critical path display also offers the possibility to tap the full potential of project simulation by setting and controlling the progress by milestones, and to react to unexpected events.

7 Conclusion and Future Research

A new simulation model based on the C3 modeling language was developed and offers project planners a suitable technique for quantitative comparisons of several alternative project structures. The influences of persons as well as tools were investigated in the first simulation runs. These experiments produced satisfactory results, as stated by experts of different leading chemical engineering companies. However, for further validation additional an empirical survey and further extensions are planned in close cooperation with enterprises. Furthermore, the correlations of individual factors are empirically calculated through the modeling of several example processes to ensure a transfer of the realizations to planned work processes.

Acknowledgements

The research was funded by the German Research Foundation according to the Collaborative Research Center no. 476, Improve.

References

[Browning et al. 2000] Browning, T.R.; and S.T. Eppinger (2000) *Modelling the Impact of Process Architecture on Cost and Schedule Risk in Product Development.* Massachusetts Institute of Technology, Sloan School of Management, Working Paper No. 4050, Cambridge, MA.

[Bortz 1999] Bortz, J. (1999) *Statistik für Sozialwissenschaftler.* Springer, Berlin, 5. ed.

[Cho et al. 2001] Cho, S.-H.; and D. Eppinger (2001) *Product Development Process Modeling Using Advanced Simulation.* In "Proceedings of DETC'01, ASME 2001 Design Engineering Technical Conferences and Computers and Information in Engineering Conference", 9-12 September 2001, Pittsburgh, PA.

[Cho et al. 2005] Cho, S.-H.; and D. Eppinger (2005) *A Simulation-Based Process Model for Managing Complex Design Projects.* In IEEE Transactions on Engineering Management, 52, 3, S. 316-328.

[Christiansen 1993] Christiansen, T. (1993) *Modeling Efficiency and Effectiveness of Coordination.* In Engineering Design Teams, PhD thesis, Stanford University, Palo Alto, CA,USA.

[Cohen 1992] Cohen, G. (1992) *The Virtual Design Team: An Object-Oriented Model of Information Sharing in Project Teams.* Ph.D. Thesis, Stanford University, Palo Alto, CA, 1992.

[Daalen et al. 1999] Daalen, C.E.V.; Thissen, W.A.H.; and A. Verbraeck (1999) *Methods for the modeling and analysis of alternatives.* in W.B. Rouse (Ed.), Handbook of Systems Engineering and Management, Wiley, New York, 1999, p. 1236.

[Eggersmann et al. 2001] Eggersmann, M.; Schneider, R.; and W. Marquardt (2001) *Modeling Work Processes in Chemical Engineering - from recording to supporting.* Technical report LPT-2001-31.

[Eimer 1978] Eimer, E. (1978) *Varianzanalyse.* Stuttgart/Berlin/Köln/ Mainz.

[Eggersmann 2004] Eggersmann, M. (2004) *Analysis and Support of Work Processes Within Chemical Engineering Design Processes.* Fortschritt-Berichte VDI, Nr. 840, VDI-Verlag, Düsseldorf.

[Gil et al. 2001] Gil, N.; Tommelein, I.D.; and R. Kirkendall (2001) *Modeling Design Development Processes in Unpredictable Environments.* In Proc. 2001 Winter Simulation Conference. Invited Paper in the Session "Extreme Simulation: Modeling Highly-Complex and Large-Scale Systems", `http://www.informs-sim.org/wsc01papers/067.PDF`, [Stand 26.01.2006].

[Gröger 2006] Gröger, M. (2006) *Wertschöpfungspotenzial Projektmanagement.* In REFA-Nachrichten 1/2006 (ISSN 0033-6874), 4-7.

[Hays 1973] Hays, W.L. (1973) *Statistics for the social sciences.* New York: Holt Rinehart, and Winston.

[Jin et al. 1996] Jin, Y.; and R. Levitt (1996) *The Virtual Design Team: A Computational Model of Project Organizations.* Computational and Mathematical Organization Theory 2:3, 171-195.

[Kausch et al. 2007] Kausch, B.; Grandt, M.; and C. Schlick (2007) *Activity-based Optimization of Cooperative Development Processes in Chemical Engineering.* In SCSC 2007 Summer Computer Simulation Conference, 15-18 July 2007, San Diego.

[Killich et al. 1999] Killich, S.; Luczak, H.; Schlick, C.; Weissenbach, M.; Wiedenmaier, S. and J. Ziegler (1999) *Task modelling for cooperative work.* In Behaviour & Information Technology, Hampshire, 18 5, S. 325-338.

[Kummer et al. 2004] Kummer, O.; Wienberg, F.; Duvigneau, M.; Schumacher, J.; Köhler, M.; Moldt, D.; Rölke, H.; and R. Valk (2004) *An Extensible Editor and Simulation Engine for Petri Nets: Renew.* In Proceedings of Applications and Theory of Petri Nets 2004: 25th International Conference, 484-493.

[Krause et al. 2004] Krause, F.-L.; Kind, C.; and J. Voigtsberger (2004) *Adaptive Modelling and Simulation of Product Development Processes.* CIRP, STC Dn, 53/1/2004.

[Levitt et al. 1994] Levitt, R.E.; G.P. Cohen; J.C. Kuntz; C.I. Nass; T. Christiansen; and Y. Jin (1994) *The Virtual Design Team: Simulating How Organizational Structure and Information Processing Tools Affect Team Performance.* In Computational Organization Theory 1994, K.M. Carley and M.J. Prietula (Eds.). Lawrence Erlbaum Assoc., Hillsdale, N.J.

[Levitt et al. 1999] Levitt, R.; Thomson, J.; Christiansen, T.; Kunz, J.; Jin, Y.; and C. Nass (1999) *Simulating Project Work Processes and Organizations: Toward a Micro-Contingency Theory of Organizational Design.* Management Science, Informs 45:11, 1479-1495.

[Licht et al. 2004] Licht, T.; Dohmen, L.; Schmitz, P.; Schmidt, L.; and H. Luczak (2004) *Person-Centered Simulation of Product Development Process using timed stochastic colored Petri-Nets.* In Proceedings of the European simulation and Modeling Conference.

[Raupach 1999] Raupach, H.-C. (1999) *Simulation von Produktentwicklungsprozessen.* Dissertation, TU Berlin, Berlin.

[Schneider et al 2006] Schneider, N.; and B.Kausch (2006) *Simulationsgestützte Optimierung der Organisationsgestaltung in Entwicklungsprozessen.* In Innovationen für Arbeit und Organisation, Bericht zum 52. Arbeitswissenschaftlichen Kongress vom 20. - 22.3.2006 am Fraunhofer - IAO Stuttgart, Hrsg.: Gesellschaft für Arbeitswissenschaft e.V.. GfA-Press, Dortmund 2006, 431-436.

[Schneider et al. 2003] Schneider, R.; and S. Gerhards (2003) *WOMS - A Work Process Modeling Tool.* In Nagl, M., Westfechtel, B. (Hrsg.): "Modelle, Werkzeuge und Infrastrukturen zur Unterstützung von Entwicklungsprozessen", Wiley VCH, Weinheim, 375-376.

[Steidel 1994] Steidel, F. (1994) *Modellierung arbeitsteilig ausgeführter, rechnerunterstützter Konstruktionsarbeit - Möglichkeiten und Grenzen personenzentrierter Simulation.* Dissertation, TU Berlin, Berlin.

[VDI 2001] *VDI-Richtlinie VDI 3633: Simulation von Logistik-.* Materialfluss und Produktionssystemen, Dez. 2001.

[Zülch et al. 2004] Zülch, G.; Jagdev, H.; and P. Stock editors (2004) *Integrating Human Aspects in Production Management.* Springer.

Index

9-intersection model, 100

actor-oriented simulation models, 213
Adaptive Resonance Theory (ART), 87
adjustement costs, 192
Adomian decomposition, 57
Agent Analyst, 108
agent based model, 197
Agent Based Modeling and Simulation (ABMS), 109
aggregate, 152
Alzheimer's disease (AD), 178, 181
ant colony, 36
ant colony optimization, 65
artificial immune network, 75
Artificial intelligent agents, 109
artificial intelligent agents, 108
artificial neural network, 202
attractive harmonic potential, 8
automatic decision making, 31
autonomous vehicles, 201

backpropagation learning, 85
backpropagation training algorithm, 203
basic emotions, 166
behavior, 152
behavioral approach system (BAS), 180
behavioral graph, 165
behavioral inhibition system (BIS), 180
behaviour theory, 178
behavioural graph, 168
behavioural intelligence, 202
Bernouilli equation, 8
bifurcation, 52
biocenosis, 142
biological theorists of emotion, 178
Biot-Savard Model, 38
Biot-Savart laws, 32
biotope, 142
bottleneck machines, 84
boundary value problems, 52
business environment, 83

Caratheodory condition, 7
catastrophe, 129
cellular automata, 108
Cellular Manufacturing (CM), 84
chalk fracturation, 122
Chalk lithostratigraphy, 120
chalk parallel fracturation indice, 123
chalk ransverse fracturation indice, 122
Chandrasekhar model, 9
chaotic dynamics, 190
chromosome, 158
cliff collapse, 118
cliff height, 120
Clinical Implications, 181
clustering, 31
Coastal piezometric slope, 124
Coastal springs occurrence, 124
coastline location, 120
cognitive agent, 173
cognitive maps, 173
cognitive theorists of emotions, 178
cognitive theory, 178
cognitive therapy, 182
colored pheromones, 36

communicating clusters, 36
communication filtering, 31
communication graph, 36
compartment, 144
competitive learning, 86
complex clustering tasks, 84
complex dynamics, 190
complex systems, 3, 140, 189
complex systems behavior, 21
complex systems dynamics, 190
complex systems modelling, 108
complex temporal-spatial behavior, 108
complexity, 129
complexity theory, 189
complicated systems, 4, 98
Componential models, 180
componential models, 179
conditioned stimulus (CS), 182
connection graph, 101
consumers, 143
control methods, 5
curvature, 42

Darboux Invariant, 42
decision making, 31, 68, 108, 165, 168
decomposers, 143
detritivors, 143
dilatation method, 104
domino effects, 130
Driver-Vehicle-Environment (DVE), 202
driving simulators, 201
dynamical interaction network, 101
dynamical systems, 190
dynamics of the productivity, 193

economic result, 193
ecosystem, 32
ecosystems, 141
ecosystems modelling, 51
edge of chaos, 190
edges, 36
Emergence of complex systems, 139
emergence of organizations, 141
emergent property, 104
emergent structures, 29
emotion, 165
emotion modelling, 178
emotional dimension, 166
emotional feedback, 169
energetic fluxes, 99
estuary, 32
Evolutive GIS Formalism, 99
expert systems, 84

facial expression, 178
Fast Multipoles Method (FMM), 33
feed-back process, 141
feedback loops, 131
fitness, 158
fluid flow, 32
food chain, 144
Forrester diagram, 132
Fourier Analysis, 23
functional analysis, 6
functional disorders, 182
Fuzzy Adaptive Resonance Theory, 89

gambling task, 165, 168
general equilibrium theory, 189
General System Theory, 140
General Systems Theory, 129
generalized Lipschitz inequality, 7
genetic algorithm, 155
Geographic Information System (GIS), 97, 108, 118
Geographical Data Base (GDB), 97
Geographical Database consistency, 101
geopolitics, 108
georeferenced data, 119
GIS updating propagation, 101
GIS: canonical operation, 101
GIS: complex semantic objects, 99
GIS: composition relations, 100
GIS: constraints between values, 100
GIS: constraints between variable, 100
GIS: geometric objects, 99
GIS: geometric primitives, 99
GIS: layers, 98, 108, 119
GIS: multiagent systems mixing, 108
GIS: raster mode, 97
GIS: semantic objects, 99
GIS: vector mode, 97
global consistency maintenance, 104
graph, 36
graph theory, 84
grid, 34
Gross Domestic Product (GDP), 190

Group Technology (GT), 83

hazard modelling, 124
hazards spatio-temporel modelling, 118
hierarchical models, 179, 180
holarchy, 142
holding costs, 25
Holistic metrics, 21
holon, 142
Homotopy Perturbation Method (HPM), 52
hybrid model, 142
hydrodynamic model, 33
hydrogeology, 123
hyperdisaster, 131

idiosyncratic shocks, 197
idiotypic network, 73
immune system, 71
individual's behavior, 191
Individual-Based Model (IBM), 108
Individual-Based Models (IBM), 140
instability, 132
Integrative Simulation Model, 215
intelligent agents, 108
interacting particles, 33
interaction network, 140
interactive networks, 108
Invariant Manifolds, 43

job shop manufacturing systems, 84

Kohonen Self Organizing Feature Maps (SOFM), 86

law-based behavior, 144
Lebesgue space, 6
Leonardo Da Vinci, 32
Lie Derivative, 42
limit cycles, 197
local consistency maintenance, 104
logic coherences, 215
Lorentz model, 45
Lotka-Volterra system, 157
Lyapounov, 7
Lyapounov method, 9

macroscopic variables, 191
major risks, 131
manifolds, 5
manufacturing companies, 83
manufacturing system, 84
Marcenkievitch-Besicovitch spaces, 6
mathematical integer programming, 84
matrix sorting, 84
mechanical representation, 4
memory model, 182
memory model: encoding, 182
memory model: episodic memory, 182
memory model: long-term memory, 182
memory model: procedural memory, 182
memory model: retrieval, 182
memory model: storage, 182
Mixing Individual-based Models and GIS, 108
motor expression, 178
multi objective scheduling problem, 65
multi-level systems, 142
multi-scale organizations, 99
Multi-scale rule-based qualitative system, 124
multiagent simulation, 31
multicast routing problem, 66
multilayer Neural Driver Agent, 202
multiscale methods, 32
multiscale simulations, 29
mutual dependance, 141

neoclassical economics, 189
Neural Driver Agent (NDA), 202
neural networks, 84
neuropsychology, 178
nonlinear ecosystems, 52
nonlinear wave equations, 52
nonlinearities, 4
numerical ants, 36

Object-oriented modeling (OOM), 108
observer, 30
OCC Model, 166
ontology, 145
organic diseases, 181
overlapping parts, 84

p-dense, 104
panic disorder, 178
Pareto front, 67
part classification, 84
part family formation, 84

part-machine grouping problem, 84
particle model, 32
pattern recognition, 84
perturbation methods, 52
Petri nets, 214
pheromone, 65
pheromones, 36
physical qualitative modelling, 124
physiological activation, 178
Poincaré, 7
predator-prey system, 56, 153
primary emotions, 166
ProActive, 151
Process Engineering, 212
process-oriented simulation models, 213
producers, 143
production costs, 192
production flow analysis (PFA), 85
production function, 191
Protégé, 145

Quantum Mechanics, 23

reductionism, 21, 141
rehabilitation, 178
reification, 32
reinforcement, 67
Repast, 108
repast, 194
research and development activities, 194
Riccati differential equation, 52
risk, 129
risk analysis, 129
risk modelling, 129

Schelling's Model, 110
segregation model, 110
self organized systems, 4
self regulation, 153
self-amplification, 36
self-organization, 99, 189
self-organized criticality, 78
Self-Organized Holarchic Open Systems (SOHOS), 142
service level, 25
sigmoid curve, 167
simple systems, 4
simulation trace, 32
Sobolev space, 6
spacial data, 97
spatial and temporal scales, 129
Stella Research Program, 133
stockouts, 25
strange attractors, 5
structure detection, 30
structure emergence, 190
summary of a simulation, 32
supervised learning, 85
supply chain management, 24
swarm, 194
system dynamics, 4
system dynamics modelling, 129
systems, 139

Task Network Model, 216
temporal coherences, 215
territorial system, 129
therapeutic applications, 178
thermodynamic representation, 4
thermodynamics, 6
topological influence area, 101
topological relations, 100
torsion, 42
turbulent flow, 9
turnovers, 25

unconditioned fear responses (URs), 182
unconditioned stimulus (US), 182
understanding, 31

Van der Pol model, 44
Virtual Design Team, 213
vortex detection, 38
vortices, 32
vorticity, 34
vulnerability, 130

water table level, 123
wolrd-wide economy, 108
Workflow simulation model, 213